On the Contexts of Things Human

An Integrative View of Brain,
Consciousness, and Freedom of Will

On the Contexts of Things Human

An Integrative View of Brain, Consciousness, and Freedom of Will

Ronald J MacGregor

University of Colorado, USA

World Scientific

NEW JERSEY • LONDON • SINGAPORE • BEIJING • SHANGHAI • HONG KONG • TAIPEI • CHENNAI

Published by

World Scientific Publishing Co. Pte. Ltd.

5 Toh Tuck Link, Singapore 596224

USA office: 27 Warren Street, Suite 401-402, Hackensack, NJ 07601

UK office: 57 Shelton Street, Covent Garden, London WC2H 9HE

British Library Cataloguing-in-Publication Data
A catalogue record for this book is available from the British Library.

ON THE CONTEXTS OF THINGS HUMAN
An Integrative View of Brain, Consciousness, and Freedom of Will

ISBN 981-256-735-6

Printed in Singapore by World Scientific Printers (S) Pte Ltd *for*

Sam, Clinton, Felicia, Jon,
Ronnie, and Terri (1962–2000)

Foreword

It is sometimes necessary, though difficult, to put oneself outside a personal orientation, regional culture, belief system, or academic field in order to better grasp its foundations, nature, outer margins, place, and shareable worth. This is the case, for example, when there is uncertainty or incompleteness in the foundations or substantive core of things at hand, in overall structural cohesion, or in corresponding sense of direction, or if there are gaps or disjointedness in its relations with other highly related or interested groups and areas. The suspicion that this has been the case for our scientific view of the brain prompted me some ten years ago to initiate this study of the foundations and wider contexts of this problem area. The overall cogency of first principles eventually adopted here and the wide scope of harmonious integrative understanding which follows in their wake seem to me the vindication of these suspicions.

This book has been some thirty years in gestation in addition to this last decade of hard labor. Blaise Pascal apologized for one his philosophical letters in Les Provinciales (ca 1657) that he hadn't time to make it shorter. Unlike Pascal, I have taken the time to make this work shorter. I have sought succinct essential statements of many complex conceptions. I believe the work is both sufficiently full and clear so as to be widely accessible despite its relation to rather complex and often difficult subject matter. An outline summary of the first section of the work and chapter seven are published in the Journal of Integrative Neuroscience, 2002 and 2004.

Integrative neuroscience is a recently coalesced field which merges theory from all of the brain sciences with a view to the deeper perspectives required by the full range and nature of the brain. This field encompasses the requirements of physical-biological foundations, expansive inclusiveness of scope across biological and psychological variables, multi-leveled hierarchical complexity, and analytical tools demanded by the brain's complexity. Integrative neuroscience embodies the future directions of theoretical neuroscience, and should provide many bridging recognitions. The first two sections of this work contain integrative material in and at the outer limits of this field. The larger interests and last section of the work speak to broader human considerations beyond science. The views in this book are my own and should not be attributed to the field or individuals in the field of integrative neuroscience, or my reviewers, many of whom may disagree with me regarding free will, the fabric or nature of the brain's inner sensibilities, or the likelihood of quantum mechanical effects in the brain.

I should like to thank my associates in the several departments of the University of Colorado, especially Max Peters, and the Rand Corporation in which I have served for their support and collegiality. I also thank at this late date my teachers at Purdue University, for their enthusiasm and classical training during my impressionable student years so many years ago, including most especially my mentor, Paul Lykoudis for his vision and inspiration, and his steady astute classical brilliance. I give special thanks to my reviewers, Roman Poznanski, Teresa McMullen, and Edwin Lewis, for their many constructive comments, both substantive and editorial. The book is considerably better in both qualities because of their efforts. I also thank James Lehman and Felicia, for pointing out literature on the neuropathology of attachment. I gratefully acknowledge reprint agreements with Imperial College Press (chapter seven), Academic Press ('stratification of variables' in chapter four and neural Darwinism in chapter five, both from MacGregor, 1993), and Mary Ann Liebert, Inc. (Figure 5 and caption, from Poznanski, 2001). I thank Millie for her interest and encouragement.

Most fundamentally, I thank the muses who have goaded and beckoned me, driven and led me.

Albuquerque, New Mexico, September, 2005

List of Figures and Tables

FIGURES*

3.1	The structure of human value	35
5.1	Self-organization of neural functions	76
6.1	Conscious autonomy and possible quantum effects	110
8.1	Neurobiological ground structure of language	142
12.1	A map of the whole things	190
14.1	Fountal streams of good and evil with ground structure	219
15.1	Contextual energization fields of personal human living	245
16.1	Areas of metaphysical uncertainty	269

TABLES

2.1	Capacities of the biological primals	19
3.1	Domains of value	38–39
4.1(a)	The hierarchical stratification of brain variables	55
4.1(b)	Topographical localization of the inner sensibilities	63–64
10.1(a)	A typology of knowing	165
11.1(a)	Representative values of philosophical ethics	177–178
11.1(b)	Placement of values of the world's religions	183–184
11.1(c)	Values of the world's religions	185
12.1(a)	Constitutive operative realms of the whole of things	194
12.1(b)	A way of placing things (third-level headings)	197–198
13.1(a)	The human condition in five compound tenets	212–213
13.1(b)	The human condition by the universal human perspective	214–215
15.1	Contextual energization fields of human living	241

*The numerals in the titles of the figures and tables identify the chapter in which the figure or table can be found.

CONTENTS

Foreword		vii
List of Figures and Tables		ix
How People Behave in Situations They Don't Understand		xiii
Ch 1 — Introduction and Overview		1
PART A — AN INTEGRATIVE VIEW OF BRAIN AND CONSCIOUS EXPERIENCE — THE INNER SENSIBILITIES		
Ch 2	An Integrative View of Brain and Conscious Experience	15
Ch 3	The Structure of Human Value	31
Ch 4	The Inner Sensibilities and Neurobiology	43
Ch 5	Consciousness and Theoretical Neuroscience	69
PART B — AN INTEGRATIVE PHILOSOPHY OF BRAIN AND CONSCIOUS EXPERIENCE		
Ch 6	Consciousness, Physics, and Neurobiology	91
Ch 7	A Functional Theory of Consciousness and its Relations in Brain	121
Ch 8	Consciousness, Inner Constructions, and Language	137
Ch 9	First Principles of Human Awareness and Apprehension	145
Ch 10	Knowing and Neurobiology	159
Ch 11	The Larger Nature and Setting of Value	169
PART C — ON THE CONTEXTS OF THINGS HUMAN		
Ch 12	A Way of Placing Things	189
Ch 13	The Universal Human Condition	199
Ch 14	Fountal Streams of Good and Evil	217
Ch 15	Contextual Energization Fields	239
Ch 16	Implications and Interpretations	247
References		271
Detailed Table of Contents		279

How People Behave In Situations They Don't Understand

If I were certain of a certain star I would whisper
what are you for star but not wish an answer
every world takes getting used to
use and getting take world after world

it's like looking in the darkness for a dark thing
or melting chips of chocolate on the stove
what meets the fingertips exists by specific accident
as if the plan is whatever happens so can't be spoiled

what is all this star   this what this is   this all
this this the smoldering smallness of my utterance burns
maybe the name of anything is ashes gnashed
in an indifferent wind ghastly haphazard wind

I say mind   I say moon   I say me   I keep losing my voice
in the space in the tree where there isn't any tree
each shadow dapples itself together as if light were something obvious
as if wishes might be granted in a place between places like that

here between sunrise and circumstance I say okay
precarious things know about being precarious things
but there is clarity between this persistence and
okay I say   where does the path disappear

Terri Jolene MacGregor, 1999

Chapter 1

Introduction and Overview

I came to feel, late in my career in theoretical neuroscience, and still do, that the conceptual constraints of our scientific view of the brain are inadequate to the field's intrinsic relevance to fundamental questions of human experience, the living of human lives, and the broader concerns of humanity. I have been dissatisfied with many overreaching claims regarding the human condition and potential which have found unwarranted widespread acceptance throughout much of academia over the last century and have now permeated pervasively throughout western culture at large. I have been driven by vague but pressing inner needs to a deeper study of the foundations and wider contexts of these larger human concerns within the partially conscious brain. I have now, after a decade of deep struggle, satisfied myself with at least an outline of the essential integrated ground structure within which these large quests can be further thrashed out.

This work, the result of that struggle, outlines a grounded, comprehensively inclusive view of man's nature, circumstances, and potential which speaks especially to the existential behavioral living of individual lives. It present a broad mainstream view of consciousness and brain which reaches to the foundations and outer reaches of the natural scope of its subject matter at and beyond the outer limits of current theoretical neuroscience. It reaches out from this scientific ground structure to the larger existential and philosophical questions regarding the fundamental nature of consciousness and its relations to brain and physical nature, freedom of will, the origin and nature of value and good and evil, our common human condition, the ambient force fields of human living, and how we know things. It discusses and interprets the central implications of these views with respect to universal human interests and problems, and to scholarly fields. It gives an integrative interpretation of important scientific foundations and principles, and relevant recent literature, focusing especially on underlying foundations and larger overall implications of this substance.

Throughout, I have attended especially to underlying determinative factors and first principles. In this, the work offers a foundational ground structure from which large issues of man, spirit, individual and collective living, morality, and the times can be fruitfully approached in terms of their essential underlying energies, constraints, and human significance. The view is grounded in the brain, conscious experience, and our ambient contexts in natural and human worlds.

1.1 Approach

All this embodies a significantly expanded scope beyond that found in recent theoretical neuroscience and the philosophy of mind. I hold that such expansions simply better reflect the natural scope of the subject matter and are necessary in order to adequately consider the natural qualities and functioning of the brain and mind. These expansions are perhaps the most fundamental premise of the work.

1.1.1 *Consciousness and brain : A multicapacitied monism*

The first expansions involve the nature of the brain itself. Advances in the brain and mind sciences promise to be perhaps even more deeply significant for humanity than have the remarkable revelations of twentieth-century theoretical physics and molecular biology. The brain with its full range of conscious experience is the vehicle of not only our abilities to see, hear, and move, our life support, creature comforts, joys and sorrows, and abilities to cope with the vagaries we encounter, but also of our understanding and knowing, our values, our aesthetic experiences, our intuitions, our senses of truth, beauty, goodness, justice--indeed the entire range of our sensibilities many of which we know only obscurely.

Yet further, our brains are the vehicle, nay--embodiment, of our very being itself. Most quintessentially, we are our conscious awareness with its full ranges of inner reflections and valuations, and its small but highly significant range of partially free individual autonomy. Of these latter consist our ultimate senses of being and our visions of our ultimate place in the scheme of things--no small things to us passing mortals.

In all this, our understanding of the brain and mind (for example, our neuroscience) influences how we see ourselves, and through this how we asses our character, make our choices and lead our lives; how we construct our cultural institutions and govern ourselves; how we direct our arts, entertainments, and scholarly fields, including those which study the brain and mind itself. Since we are a plastic, learning, open-ended species, what we think we know strongly influences what directions we pursue, what we continue to think, what we learn, indeed, what we largely are and what we may become. What we next construct grows from what we have constructed.

This higher plasticity of learning and its intimate association with our partially autonomous consciousness is the great feature of our species--that we are partially open-ended with rich capabilities for adaptive constructions and even richer capacities for deep free inner reflection and valuations, as will be discussed throughout the work.

A closer look at the relations of consciousness and the brain reveals explicit difficulties. The study of consciousness faces two bodies of observations, those of the existential experiences of consciousness itself and those regarding its relations within the brain. A theory of brain and consciousness must answer both of these perspectives. Taking the existential first, note for one thing, as any individual can readily attest by introspection, that one's consciousness contains a relatively rich multicapacitied mix of sensibilities across feelings, apprehensions, adaptive proclivities, and a range of higher valuative representations including, but by no means restricted to, rational thought. This work incorporates this rich multicapacitied sensibility within an inclusive functional interpretation of the existential-operative partially conscious brain.

The work explicitly recognizes the natural physical-experiential wholeness of our sensibilities and existence: physicality, biology, physiology, neurobiology, all conscious experience; all its physical, experiential and unconscious faces, processes, and integrations, with all their obscurities, uncertainties, and ambiguities; conscious and subconscious mental activity, emotions, aesthetic experience, our senses of continuity of conscious operative being with all its senses of vitality, adventure, humor, joi-de-vivre, melancholy, agony, transcendent integrations and peak experiences of all kinds including deepest spiritual and religious experiences. In this, the work endorses and develops a functional monism of neurobiology and conscious experience with all its intrinsic existential senses of being, open-ended reflection, imagination, and varied experiences.

The explicit recognition of conscious partial autonomy also necessitates the recognition of conscious-related qualities such as higher sensibilities and judgments as major players in brain-mind function. This work recognizes the central functional and existential role of such higher human sensibilities as aesthetic perceptions, cognition and reason, and value itself, in fluid intracooperation with animal-like biological primals, practical adaptive integrations of everyday living, and the integrated functional whole which is composed of these. It leads us to see the operation of the brain as an organic functional unity of integrated cooperation of experiential and neurobiological variables. This is the first major expansion of this work. Its development is the subject matter of Part A.

1.1.2 *The obscurities of consciousness*

The explicit integration of consciousness with the brain's neurobiology forces us to deep obscurities in the nature of the integrated brain. Most fundamental is the immediate existential sense of any human that he is one with conscious awareness and autonomously directs the living of his life from this center of awareness. This universal observation entails major difficulties with physical science. Neurobiological brain functioning must be seen in terms of constituent physiological processes all driven in accordance with physical law. This leads us into a complex functional physical structuro-physiological hierarchy. Within this hierarchy, consciously driven partial free will seems to expose itself as an apparent paradox in operating within an otherwise self-determinative, perhaps exactly deterministic, physical neurobiology[1].

Some scientists have recently interpreted brain function in terms of the chaos theory of complicated systems which are non-predictable but not random, like the weather[2]. This is a statistical systems model which characterizes the system in terms of probability functions rather than exact values and sees variations and uncertainties as unavoidable 'noise'. Some see the possibility for autonomous choices within these uncertainties and the possible grounding of consciousness in the brain in relation to this freedom of selection.

A major difficulty here is that in classical physics the unpredictability of the chaos theory of purely physical systems arises not from intrinsic non-predictability of the system itself but from our human inability to completely describe the vast numbers of variables which determine its behavioral states and trajectories. According to classical physics, all macroscopic physical systems operate in terms of exactly deterministic physical laws and will, in fact, run off their behavior along a deterministic course according to their ambient and internal forces and initial states whether we can calculate them or not. This trajectory would include the mechanics of all the noise of a classically described brain and could be observed and calculated by a sufficiently vast intelligence. Our tremendous success with predictable physical nature and technology substantiates our belief in this exact determinism. However, some biological theorists doubt the principle of exact determinism even for classical macroscopic physical systems.

The statistical systems model in itself can provide only an open door to the sources of indeterminism it stresses, but does not speak in itself to their physical nature nor relation to consciousness. It has, however, encouraged interest in experimental studies of various sources of neural noise.

There is a second major philosophical problem with the idea that the physical brain can do it all. On the one hand, most neuroscientists hold that consciousness is an emergent quality of the nervous system. Many, perhaps most, also believe that the physical brain does it all, and that partial conscious autonomy reflects the rich complexity, vast quantities of combinations and permutations, and pervasive unpredictable imperfections of its complex neurobiological systems, and the unique idiosyncrasies of individual learning, experience, and local circumstances. In this view all our preferences, choices, even our sense of whim--for example, the ability to move or not move one's finger at any given instant--would be seen as ultimately servant to the global mix of neurobiological signals and states. In such a view, though, there is no need for consciousness-- it is merely a deluded rider in a neurochemical soup. Its autonomy is illusory, a false autonomy. Philosophers have used the word epiphenomenon to describe this case. I would say, further, that the observed existence of an epiphenomenal consciousness violates a fundamental tenet of the construction of scientific theory--that of the minimal necessary hypothesis (Ockham's razor). The logic of science itself would imply that if consciousness is there, then it must do something beyond what even a determinative but non-deterministic brain does without it.

Overall, I hold that the arguments that the traditional physical brain does not do it all do not depend upon whether the physical brain is exactly deterministic or only determinative with some unavoidable uncertainties. Rather, I say that the conceptual difficulties highlighted by considerations of partial autonomy acting within a determinative brain identify the core of the mind-body question as one of energization--the ability to exert a force, to act as a cause of motion or change. If consciously driven autonomy is a real autonomy it must make selective effects on the brain's neurobiology beyond what the physical brain would have done without it. Its clarification resides in the relationship between partial autonomy and physical force. I hold further that the existence of partial conscious autonomy is too centrally pervasive in human life to be denied. I say rather that its existence requires the existence of a non-traditional influence in the brain's neurobiology because there is no place for such autonomous action within traditional physics.

Most physicists hold that the classical deterministic rules apply to all macroscopic systems with characteristic operating distances above a little less than 10^{-9} meters (a nanometer). This is more than a thousand times the typical micron-level operating distance of most neural processes. However, there are both experimental and theoretical reasons to suspect that microscopic quantum-mechanical effects may occur in the brain's neurobiology or metabolism and that these may influence macroscopic neural variables, as is described in Chapter 6. This is important especially because it introduces both non-deterministic uncertainties and non-traditional mechanisms into physical processes.

Chapter 6 discusses all these issues more fully within a larger search of the role of current theoretical physics regarding consciousness and free will in the brain's neurobiology. That chapter describes and discusses the structural basis for possible quantum mechanical effects in the brain's neurobiology, and identifies several candidate possibilities. If consciousness and autonomy can be resolved with brain operations by quantum mechanics, all well and good, but there are very important fine points to clarify and a great deal to establish well beyond the simple pointing out of a possibility.

This work, then, recognizes a partially autonomous functioning of consciousness in brain operations, its central involvement in directing behavioral living, and the need to address the deep difficulties associated with this supposition. This is a second major expansion of this work. In deference to these difficulties, this work's functional monism endorses neither a substantive monism of identity of experiential and neural variables nor a mind-body dualism. This is developed in Part B.

1.1.3 *The grand unifying vision of science*

This work holds a deep affinity with the grand unifying vision of science, especially science's pursuit of an harmonious synthesis across the three major realms of: physical science, biology, and brain-consciousness-behavior. Each of these levels can be seen to house its own self-relevant patterns of activity, organization, and energization. Each level has its obscurities of origin and essential nature. The central remaining obscurities in this grand synthesis of science are at the origins and interfaces of these three levels: the origin of the universe, the presumed natural origin of the DNA molecule within the physical forces of nature, and the fundamental nature of consciousness and its linkages in the brain. These may be entry points to watershed convolutions in our scientific and perhaps common sense understandings of the whole of things.

The last section of the book presents structured views of the ambient contexts and energization fields which govern human living. This is the third major expansion in this work.

1.2 Overview

The first section of the book outlines an integrated view of the conscious brain as a cooperative group of three sets of functional capacities including: biological primals, adaptive integrations, and higher valuative representations, the last including generalizations of aesthetic perceptions, apprehension-cognition, and value. These capacities are described in chapters two and three and discussed with regard to neurobiological and evolutionary theory in Chapter 4. Consciousness is discussed with regard to current neuroscience and underlying neurobiology in Chapter 5.

The second section discusses the relation of the fundamental nature of autonomously conscious brain in terms of modern physics as indicated above, and offers, in chapter seven a theory of consciousness as the director of behavioral living in intimate association with the ongoing cultivation and utilization of vast systems of patterned inner constructions. Chaper 8 provides a description of language as a cognitive inner construction which mediates between consciousness and the fuller neurobiology of its various contents. The next two chapters offer an overall theory of knowing, grounded in first principles of pre-analytic apprehension and relations to current epistemology, and culminating in the representation of cognitive elements and analytic operations in the neural circuits of the cerebral cortex. The larger operational nature of value, its linkages with cognitive elements as valenced thoughts, and its relations to ethics, wisdom, and religious values, are indicated in chapter eleven.

The last section offers systematically derived integratively structured views of: a universal human perspective of the whole of things as experienced directly by every individual human, the universal human condition, good and evil seen as inherent in the human inner sensibilities, and the dozen-odd fields of inner and external influence on the common daily life of humans.

The last chapter summarizes implications regarding the work for science, philosophy, individual and collective human living, and religious and spiritual interests.

Overall, the book can be seen as defining the functional structure of the brain in terms which are deeply consistent with both neurobiology and experience, and which give ground structure for common and philosophical considerations of our human nature, condition, and potential. In this, the book defines a neurobiological basis of essential higher sensibilities and consciousness in the brain's neurobiology in relationship to human living and considers its implications.

In a very broad sense, the work may be seen as taking us to an outer edge of the knowing of things at the depths of our common human-centered perspective. It identifies the human-centered platform from which our knowledge will continue to unfold and grow.

One important result is the perspective to see most of the limited proscriptions of post-modern thinking as overwrought inferences from half-truths, which collectively and falsely severely overreduce the fuller human potential indicated by this work. This book leads us, rather, to see the fuller, open scope of our multiple inner capacities and rich plastic nature, and to recognize our large open-ended potential and freedoms of choice and growth, both individually and collectively.

The work as a whole might be seen as an expanded twenty-first-century adaptation of the four questions posed for the study of the brain and mind by William James a century ago[3]: What is the nature of the mind and brain? What are its surrounding contextual circumstances and place? How do we know these things? What are the implications of all of this?

No single work of such scope can be complete and richly detailed within itself. The work itself is an outline--an outline which identifies and places the causal foundations and central areas of its subject.

Superscripted numerals refer to notes listed by chapter on pages 251-255 which include comments and citations to literature given in a separate alphabetical listing. A detailed table of contents serves as a cogently organized index to topics considered in the text.

1.3 Personal Proclivities

Finally, I have brought to the work, along with a compelling intrinsic interest in the subject matter, several personal inclinations, including: a need for comprehensiveness-- the work responds to the truth in the half-truth that 'one must know everything before knowing anything'; a desire for universality-- the work has sought not a particular philosophy but the ground structure from which all such philosophies must flow--the goose not the egg; a need for fundamental directive influences--the work has sought, wherever possible, the identification of the underlying energizations of things, even at levels of abstraction far beyond any clear relationships to underlying physical law as such. This last is in accordance with the methodology and spirit of science generally, and the grand integrative vision of science indicated above.

1.4 Terminology

A broad integrative work such as this which attempts to probe both deeply and comprehensively across a vast range of material, in, across, and outside of disciplines and their margins, necessarily requires the construction of some new terminology. Throughout this work, I have adopted terminology as closely consistent with my intended meanings and with existing English meanings as has seemed right. Meanings are indicated at the first usages of terms and should be or become clear to readers through natural usage in the text. Where it has seemed important, I have stated definitions. I have taken 'inner sensibilities' as a global inclusive term for the capacities of the integrated brain and consciousness because it continually suggests the relation with conscious experience. I describe the inner sensibilities in terms of its 'capacities' in an all-inclusive sense, but also with special emphasis on their operational and functional dimension. Inner sensibilities, 'mentoexperiential', and 'mentospiritual' are defined in the next paragraph, as are my uses of 'spirit' and 'spiritual'. I use the term 'energization' rather than, say, 'force' because its depth, power, and easy general applicability. I take it in both specific physical or more general interpretational senses of the ability to exert a force--to bring about motion or change, to make things happen. Its meaning in any particular instance usually can be readily discerned. 'Primal' means 'first, original, primeval', 'of first importance, fundamental'. It is usually used as an adjective, but also flows easily from and into one's thinking as a noun. I have used it throughout as either adjective or noun because it conveys precisely, succinctly, and pleasingly as no other word can my repeatedly needed meaning. I have used the word 'fountal' in the sense of: 'of or pertaining to a fount or source'. I have used the word 'valenced', as equivalent and interchangeable with 'charged' in the sense of being endowed with an energizing quality. I use it particularly in the term 'valenced thought' for a thought associated with a distinct value beyond the cognitive dimension in itself. This is similar to the use of the word valence in chemistry to indicate the positive or negative electrical polarities which characterize ions (molecules which have gained or lost electrons so as to disturb their root electric neutrality).

I use 'deterministic' in the widely recognized sense of lawful exact predictability, but contrast it with 'determinative' as a weaker, less definite sense of 'giving direction or tendency to', or 'serving to cause, direct, or impel'. I use the word 'autonomy' always as an autonomy of consciousness and equivalent to free will, and take both to imply a freedom of selective initiation of action beyond all other determinative or deterministic processes of the physical brain in itself. This is quite different from the use of the word 'autonomy' applied in robotics, for example. Human free will (and autonomy) is partial in that it is inescapably accompanied by circumstantial features given by the external ambient environments and much other brain activity which have their own physical origins and momentum. The brain is seen here as partially conscious because we assume that consciousness is associated with some brain activity and that much of brain activity is not conscious. These matters are discussed in chapters four through seven.

I will use the following terminology throughout the book to mark out and maintain the neuroscientific integrity and relevance of the central parts of this work. I will use the term 'plastic' as it is universally recognized in neuroscience: as functiono-structural alterations in the nervous system which underlie any kind of accomodation, adaptation, conditioning, or learning. These include such things as selective alterations of synaptic interconnective strengths in cognitive neural networks, or increases or decreases in neuronal responsivenes to sustained input. I define the term inner sensibilities as the integrated unitary brain and consciousness, especially referring to all the explicitly verifiable neurobiological variables and processes of the brain and also all those of consciousness recognized by introspection. I mean by the terms 'mentoexperiential' and 'mentospiritual' all those variables and processes centrally associated with conscious experience and its associated unconscious processes. The terms simply recognize both the distinguishability of conscious-related variables from neurobiological variables as phenomena and the need for clarification of their underlying relations within or with neurobiology. The terms do not necessarily imply nor endorse an acceptance of dualism, and are indeed in principle compatible with the strictest monism, as well as with alternative middle grounds, as discussed in chapter six. All variables acceptable by these terms are associated with some regions of verifiability in conscious experience. I make no assumptions regarding their obscure outer margins.

At a broader level, however, I intend the terms mentoexperiential and mentospiritual to also serve as inclusive higher level indicators of all the ambiguities, uncertainties, and different points of view implicit to the area so as to encourage both the continuing explicit integrative clarification of these in close intercomparison with neuroscientific views and the effective communication with various different viewpoints in or out of science. I generally will avoid the terms 'spirit' or 'spiritual' except in relationship to common general usage such as: the principles of conscious life or vitality in humans with no supernatural implications, or when considering religious or mystical interests, eternal life, and the like, or God.

I sometimes use a pedagogical 'we' which invites readers to join me for the moment in considering a given direction of thought. Other times 'we' may refer to a universal human quality.

Endnotes

1. Terms used frequently in this book, such as autonomy, free-will, and determinism, are defined in the "Terminology' section on pages 10-12.
2. Korn and Faure2003
3. James1890

PART A

AN INTEGRATIVE VIEW OF BRAIN AND CONSCIOUS EXPERIENCE — THE INNER SENSIBILITIES

Chapter 2

An Integrative View of Brain and Conscious Experience

This work rests on the presumption that human experience and internal control are mediated by an organic functional integration of consciousness and the brain, of existential experience and neurobiology, which view, in itself, is consonant with the thinking of most neuroscientists and the public at large. I use the term 'inner sensibilities' as the integrated unitary brain and consciousness, so as to especially refer to all variables of conscious experience verified by introspection, as well as to all the explicitly verifiable neurobiological variables and processes of the brain. The explicit recognition of function given here as an overriding dimension of brain-mind organization is important. It relates closely to evolutionary selection which is taken here to have guided the structural development of the brain. It relates directly to the practicalities of human existence. It relates directly to our quintessentially important sense of personal autonomy. It underscores and helps identify, as we will describe below, the nature and significance of our higher sensibilities as operative within and beyond practical adaptive functioning. This work defines functionalism in an operative sense only. The work does not take a stand regarding philosophical functionalism, a view in the philosophy of mind which relates consciousness to brain activity according to causality, except to insist on the partially autonomous action of consciousness in or on brain activity.

This chapter introduces a fundamental theoretical view of the functional capacities of the inner sensibilities, built on this presumption, which is consistent with and encompasses both the collective neursocientific knowledge of the brain and the common everyday experience and behavioral living of individual humans. I will try to persuade you here and throughout the book, of the essential cogent validity of this view. I personally see its qualities as easily recognizable in both individual introspection and external manifestations, and the view itself ultimately justified by the wide scope of harmonious integrative understanding which follows in its wake. This view is a high-level top-down conceptual model and many of its central qualities are necessarily embedded in the obscure complexity of higher-level brain integration. This is an inescapable quality of the subject matter rather than of this model in itself. Nonetheless, the model's components are related in this book to brain neurobiology to the extent possible and appropriate for this work, and are ultimately testable. Relevant literature is discussed in subsequent chapters, and the model's implications are particularly pursued for brain-consciousness relations and analytical knowing. The work's views of motivation are introduced later in this chapter and pursued in later chapters.

About a century ago, Hughlings Jackson, an English psychiatrist, offered the theory that in the course of evolution, the vertebrate brain attained several stable levels of overall operative organization, which he labeled 'levels of construction'[1]. In progressively higher forms, the lower levels with their functional capacities largely intact, are subsumed within the more recent higher-order level of organizational construction. This large-scale principle reflects the widely held assumption (still current) that reptiles and mammals (as well as some other groups within and outside these) each stem from single common ancestors[2]. Jackson is also asserting that some organized electro-chemical patterns themselves, in at least the vertebrate nervous system, determine a certain self-sustaining tenacity of functional construction somewhat beyond genetic prescription in itself, which latter may vary in important particulars across species as well as across individuals within larger groupings of either. Jackson's principle is upheld by subsequent estimates of the brain's functional topography (see Chapter 4). In mid twentieth century, Paul MacLean, a New York psychiatrist, interpreted the structuro-functional regions of the vertebrate brain into three levels on this basis as: a brain-stem reptilian brain (appetitive), a limbic (mid-brain) mammalian brain (emotional), and a cortical human brain (cognitive)[3].

This work offers an alternative formulation of Jackson's principle which seems more in keeping with the observable operational structure of human as well as vertebrate life. In this we see the human brain as composed of three hierarchically nested functional realms, which can be roughly indicated as: animal-like, adaptive integrational, and valuative representational sensibilities. These labels are abstract conceptualizations, yet are intended to indicate sets of naturally occurring functional systems and subsystems within the brain and mind. Many are already clearly identified and located in the brain. Others are obscure, diffuse, or global, highly difficult to localize; others are only roughly identifiable or as yet unknown. Nonetheless, this work holds that this grouping best recognizes our human hierarchical functional structure. These three levels are seen here as operating cooperatively in the full human brain and mind much as the progressive levels of construction in the Jacksonian sense, even though their localizations in the brain are various, diffuse, global, and not fully known.

The animal-like capacities include rudiments of body and life support, and appetitive control much like those common through much of the animal kingdom. Our adaptive behavior may be seen as fundamentally similar to that of higher mammals, but much more richly diverse and extensive. Our higher valuative representation capacities are (almost) uniquely human (see Chapter 4), and are our most useful and most consciously meaningful capacities.

Each of these three functional capacities is in turn richly multicapacitied or multidimensioned in itself; the whole ranges widely across the most rudimentary bodily appetites to the most ethereal realms of aesthetic experience, understanding, and senses of ultimate meaning and significance. The whole is a physical-experiential functional organic whole which ranges across rudimentary physiological structures which monitor bodily functions (well-understood in terms of physical process and physical laws), much highly complex internal representative functioning and adaptive integrations (obscurely but surely neural based), and our most precious senses of personal existence and autonomy (whose relations to neurobiology are less clear).

2.1 Individual Components of the Inner Sensibilities

The animal-like, adaptive, and valuative representational capacities, are labeled, respectively, 'biological primals', 'adaptive integration', and 'valuative representation'. All of these capacities and especially their subcomponents overlap and interpenetrate, sometimes extensively. The biological primals are largely genetically prescribed. Both the adaptive and valuation capacities, however, are highly dependent on training, condition, and learning incurred within individual experience, and are variably modifiable throughout life. This vitally important quality of learning and modifiability is referred to as 'plasticity'. Both adaptive and valuative capacities use a mix of genetic prescription and plasticity; the ratio being more highly tilted towards the plastic in the valuative. This section identifies the capacities more specifically; their anatomical locations in the brain are indicated in chapter four.

2.1.1 *Capacities of the biological primals*

We take the biological primals to be operative capacities associated inclusively with all those neurobiological structures, physiological processes, and mentoexperiential experiences associated with basic biological needs arising in our foundational rootedness within life on earth. These capacities can be grouped according to: life support, animal vitality, and health; appetites and rudimentary appetitive behaviors; sensations, rudimentary movements and expressions; emotions (polarizations and well-being) and feelings; and instinctual overall regulation and archetypical internal templates. These primal biological capacities are a central, but not sole, source of individual energization and motivation. They are largely genetically predetermined, hard-wired and fixed in the neurobiological structures and processes of the brain and body. They are highly developed in humans, but share most attributes with other mammals and some with a wide range of lower forms. The fuller range of these capacities is indicated in Table 2.1.

Table 2.1. Capacities of the biological primals.

Life-Support, Health-Vitality
{ respiration, cardiovascular, body temperature, chemical integrity, bodily homeostasis }

Appetites
{nutrition, creature comforts, ... drives ... sexual, hunger, thirst ... desire ... passion... secondary: toys, baubles, status symbols ... }

Sensations-Expressions
{... sensory ... pleasure, pain ... movements ... expressions...}

Emotions
{ polarizations: malice, hatred, rage, anger, jealousy, envy, disgust , dislike, aversion, neutrality, curiosity, confidence, trust, attraction, like, love .. }
{ well-being: anguishm grief ... fear, anxiety ... sorrow ... melancholy ... mood ...contentment, pleasure ... excitement, thrill, happiness, joy, ecstasy ... }

InstinctuaL Behavioral Regulations and Reflexes
{sensorimotor reflexes ... arousal, fight-flight, stress ... fixed action patterns... hunting-gathering ... acquisitiveness ...territoriality ...place ... nurturing.. empathy, aversion, intrinsic bonds and tensions...kindredness, alienations...}

The operations of body and life support are rooted in the neurobiology of peripheral, spinal, and brain stem regions. Emotions are associated largely with limbic and mid-brain centers, appetites with mid-brain and brain stem regions. Of the three main sensibility components, the biological primals are the most animal-like and the most fully determined by genetics.

2.1.2 *Capacities of adaptive integration*

The capacities and processes of adaptive integration establish and oversee behavior patterns appropriate to given situations and environments. This includes planning, anticipation, the ordering of behavioral elements in time, and the construction of behavioral patterns appropriate to both natural environments and sociocultural situations.

Adaptive integration operates with a mix of highly plastic learned patterns often built upon genetically determined basic structures. Much of this operation makes use of basic plastic expressive modules of movement, socialization, language, and various skills and habits. These processes of adaptive integration are determined by a combination of genetic prescription, external experience, and one's partially free directive efforts. This mixture of partial freedom and constraint (internal and external) is endemic to adaptive integration.

More broadly, the main operational dimensions of the adaptive integration capacity can be taken to include: plastic expressive modules; internal and interfacial cohesion and integration; autonomous adaptive direction and the particulars of one's present circumstances; degrees of personal individuality--individuation; and universal and transcendent levels of integration. Plasticity is pervasive in these capacities.

We can see adaptive behavior in many lower animals, and evidence of more in early primitive periods of human history. The practical living of individual human life within any historical period is governed by these processes of adaptive integration and oversight--this is largely a domain of one's particulars of interconnections, circumstances, and interengagements with the outside world.

These capacities are highly plastic and malleable, adaptive to external circumstances. They are more highly developed in humans than in any other life forms. They operate with a mix of conscious mentoexperiential control and subconscious automatic systemic operation. They are associated with neurobiological processes of the spinal cord, brain-stem, mid-brain, and frontal regions of the cerebral cortex, inclusive of all sensory, motor, and expressive systems.

2.1.3 *Capacities of valuative representation*

Valuative representation houses three central operational capacities: perceptual aesthetic evaluation; a complex of cognitive-like and cognitive apprehension which includes understanding, reason, knowing, and truth; and valuation in itself--goodness and badness. Each capacity defines its own unique dimension of judgment and corresponding constellation of operational processes and rules. The valuative representation capacity can operate in terms of its individual components or their global integration across the entire inner sensibilities. The operational forms of its three main capacities are genetically prescribed, but they all operate with broad use of plastic content, much of this latter associated particularly with the apprehension capacity.

2.1.3.1 *Generalized aesthetic perception*

The capacity of aesthetic perception registers an overall evaluative response to a situation, event, or object, usually in the here-now present, usually tangible, and in conjunction with one's overall internal resonant responses to this engagement. Perceptual aesthetics evaluates internal perceptual signals and apprehensions of objects and events of external reality according to its own intrinsic independent dimension of operations, qualities, and judgment--inclusive of overall internal sensibility responses and perhaps internal templates. It is a generalized foundation of the aesthetic sense, and inclusive of all the central characteristics of the aesthetic dimension. We see it as distinguished by a tendency to characterize the essential nature of objects or situations relative to one's interests, according to one's global internal responses to these. Aesthetic perception may reflect a primordial evaluating of objects of external reality for essence or authenticity, either as a primitive aesthetic-apprehension-integrative system in itself or in conjunction with such. It may be the seat of a primitive kind or quality of apprehension which we call intuition. In this general capacity, aesthetic perception may be characterized as operationally contributing to the question 'what is this?'

Some main qualities associated with perceptual aesthetics are beauty, proportion, balance, any number of sensory particulars in all modes (mostly genetically determined--shapes, textures, arrangements, harmony, melody, aroma, taste, and so on). Aesthetic perception includes inherent sensitivities to a number of forms and qualities of our ambient natural and interpersonal environments, many of which may reflect genetically-prescribed archetypes--bodily forms, landscapes, and the like. A prominent operative mechanism of the aesthetic dimension is a 'just-so ness' of comparison among inner neurobiological representations.

2.1.3.2 *Apprehension-cognition*

The capacity of apprehension is our central system of valuative representation. It consists of a wide array of subrational and rational processes of apprehension, association, cognition, understanding, knowing, reason, interrelational mapping, and the like. Its valuative dimension is truth. Its rational capacities are rooted in analytic processing. Interrelational mapping may be seen as a generic spatial or spatial-like capacity for arranging, and includes sequences, hierarchies, and other meaningful configurations. It operates in close intercooperation with perceptual aesthetics.

Apprehension involves the formation of plastic internal representations of various objects and qualities, and unlimited constructed abstractions. The most primitive are simple learned associations reflecting plastically strengthened neuroelectric couplings. Interrelational mapping may utilize spatially-related circuitry of deep regions of posterior visual cortex. The more advanced rational capacities may likely operate in terms of coordinated dynamic neuroelectric patterns in neural networks of the cerebral cortex.

Apprehension can create internal constructions without limit to represent things, their attributes, and their internal operations. Such constructions can be hewn to model external reality and help anticipate and guide our actions within it. Alternatively, such constructions can be seen as imaginary creations, of interest in themselves. Cognitive representation can be used to construct chains of premises and conclusions in conjunction with its own and the other judgments to produce what we call reason or understanding. It defines for us the dimension of truth, and gives us tools with which to pursue it. These latter produce human cognitive qualities of logic, rationality, coherence or correspondence and the like. Many of these in turn, may be seen to ultimately reflect comparisons or interrelations across representations.

Cognitive representation can be used practically to imagine and construct all sorts of artifacts to enhance human living in either natural or man-made environments, which latter it has constructed as well. In this it cooperates intimately with the adaptive integration capacities. Our cities, cultures, and modern civilizations reflect largely both adaptive and freer valuative open-ended plastic inner constructions.

Apprehension, with its cognition and reason, provides a full, self-contained theater of quintessential human activity in itself, serves the practical needs of adaptive integration, provides images and representations for perceptual aesthetics, and cooperates with the value capacity to produce value-weighted cognitive elements, labeled 'valenced thoughts'. Its higher capacities in particular are pervasively plastic. Apprehension operates in and out of consciousness. It consists of a multiplicity of ways of processing representations within the full inner sensibilities, and higher-order composites of such representations. Its operational processes include simple interconnective associations, spatial-like arrangements and mapping, symbolic representations, analytic distinctions, deductions, inferences, integrative coherences, and likely others less clear.

Cognition and reason serve the inner sensibilities by the production and manipulation of internal representations. This general capacity operates freely unto itself, with no necessary relation to externals, and with the ability to create new, modified, or composite representations unto itself. These representations may include those particularly representative of the objects, nature, and processes of external reality. Making models of external reality and its processes which can be experimentally manipulated is one of its most practically useful capacities, particularly seen from the perspectives of survival and autonomous adaptive direction. The processes of comparison assess for coherences or correspondences of representations which latter are widely held as the foundation of the human sense of rational analytic truth. These may be based on a small group of neurobiological processes involved in match-mismatch operations. In this, cognition-reason may be characterized as operationally serving 'what is true?

2.1.3.3 *Value, good, and bad*

The third main valuative representation capacity is the dimension of valuation along an axis of good and bad. Value intrinsically valuates internal signals of all kinds in terms of significance as a dimension of good-bad according to its own self-defined sense of value. It is widely and inclusively sensitive to one's sense of the overall well-being of one's self and the whole of things, inclusive of all one's sensibilities, external interrelations, circumstances and contexts, inclusive of all one's apprehensions, reflections, anticipations, and universal valenced thoughts, and inclusive of the objective qualities of the functional capacities themselves.

Value is characterized in chapter three in three main classes:--intrinsic, sociopersonal (mixed and associative), and valenced thoughts--which last are formed in conjunction with the apprehension capacity. Intrinsic values are associated directly with mostly genetically-prescribed processes--including capacities of body physiology, biological primals, and intrinsic senses of perceptual aesthetics and cognition-reason in themselves. Value also includes intrinsic global manifestations unique to itself, perhaps diffused widely throughout the inner sensibilities. Valenced thoughts are principles and complexes of the cognitive-reason capacity weighted or, as we shall say, charged or valenced by linkages to intrinsic values.

The value capacity is the embodiment of value in humans relative to the state and events of the inner sensibilities and ambient external reality. It is the essential source of ultimate significance in human inner experience and of self-direction in individual and collective humanity. It is fundamentally global and diffusely pervasive throughout the sensibilities, and is also intrinsically represented within most sensibility capacities. Value is an intrinsic independent dimension of operations, qualities, and judgment along an axis of good-bad. In this, value may be characterized as operationally serving 'what is good?'. Many human values are associated with biological rudiments (survival, hunger, appetites, place, territory) and share specific values and the dimension of value with many other life forms. An equally wide range of higher representational values, mostly aesthetic, cognitive, or plastic or both seem to be uniquely human (valenced thoughts, beauty, grace, morality, ethics, and the like).

2.1.4 *The valuative representations as a whole*

The valuative representation capacity may operate as a coherent whole when serving as valuative guide for adaptive integrations in personal living, and when making overall judgments and decisions. Justice is a concept that comes close to embodying the confluence of all three dimensions aesthetics, reason, and value. Wisdom and religious beliefs can be seen as mixes of value and reason. Many other attributes of human judgment reflect such confluences of judgment. Its three individual capacities are easily and clearly separable, both in the definitions given above and in experience. There is nonetheless sometimes a sense of a kind of 'truth' at the depths of some aesthetic judgments. This may be associated with a primary sense of essence, authenticity, or intuition useful in a primitive aesthetic-cognitive-value system, as suggested in the description of perceptual aesthetics above.

There is also some kind of beauty at the depths of some truths. Probably this is associated with an intrinsic aesthetic appreciation of an overall harmony of cognitive complexes, or of a larger whole of which the truth is part. The intrinsic sense of truth itself may be beautiful. Such valuations are valid in themselves as reflections of the aesthetic dimension, but are quite separate from the truth in itself. Many truths are not beautiful; many beautiful things are not true. The fuller ranges of perceptual aesthetics and apprehension-cognition both contain murky peripheral margins which may well overlap, and may exhibit some remnants of primordial functional conjointness, as we have indicated. Value cooperates with reason and may overlap with it and aesthetics, but is easily distinguishable from both. This will be made clearer in chapters three and eleven.

All three dimensions of aesthetics, reason, and value are vital in both individual and collective human living. All three are limited, often uncertain, and often judge differently in different individuals and groups. All three attain to senses of their own sovereignty. Any may become temporarily superordinate. Perhaps they interrelate generally in this ultimate order: value (deepest), reason (most testable), aesthetics (least superordinate). Yet, individuals vary deeply in this.

2.2 Motivation and Inner Directive Causes, Energization

This chapter has so far outlined the operative capacities of the human inner sensibilities. It now needs to identify their inner directive causes--drives, motivations, wants, and volitions--their energizations. The motivational-energizational structure of the human inner sensibilities consists of a complex hierarchy of a half-dozen quasi-independent evolutionarily-related subsystems. These may be seen to operate in terms of mostly temporary local dynamic activity patterns or disturbances, many of which we may see as goads to emphasize their motivational nature. Physiologically these may be standing patterns of neuroelectric signals in recurrent networks of neurons, or neurochemical accumulations or imbalances. Structurally, they may be constructed or guided according to fixed inner patterns such as archetypes or learned plastic 'inner constructions'. An archetype is a relatively fixed genetically-prescribed or early-learned inherent set of anatomical particulars in the brain which under appropriate conditions may produce specific patterns of physiological activity (coordinated patterns of neuronal firings in a neural network or in feed-forward associative connection, coordinated patterns of neurochemical secretions or targetings, and the like) representative of something. An 'inner construction' is a learned and more modifiable set of such patterned anatomical particulars with a wider and richer range of development and differentiation.

Main motivational-energizational systems are[4]: primitive sensorimotor reflexes; arousal systems mediating startle, arousal, attention, fight or flight; higher ethological instinct patterns which utilize sequences of coupled trigger-stimuli and fixed action patterns; appetitive systems which recognize incentive stimuli and action-contingent situations; affective systems which utilize value-weighting and significance recognition; adaptations and higher valuative-representations both of which utilize learned plastic inner constructions; and overall direction and guidance by partially autonomous consciousness, perhaps operating by feedback according to a vast system of inner constructions and a continuously-varying set of flags indicating current contingencies (see Chapter 7). Physiological systems: operate according to wonderfully delicate diffusions and fluxes of molecules and ions through body fluid volumes and across cell membranes, driven by concentration gradients and carried by pressure-driven fluids; act by chemical reactions; are governed by complex electro-chemical feedback control. Neurobiological systems use these processes and add the dimensions of extremely finely sculptured elements and patterns of neuroelectric signaling and neurochemical transmissions (see Chapter 5).

The densely interconnected neural networks of the nervous system, by means of their neuroelectrical and neurochemical signaling, possess remarkable natural intrinsic self-organization, inner logic of operation, and natural internal momentum, much of which may go on with no involvement of consciousness. In humans especially, however, the highest levels of control seem mediated by conscious direction, and these make very large use of learned plastic inner constructions. Maslow, for example, stresses an existential distinction between deficiency motivation and growth motivation[5].

For humans, we take this entire array of motivational energizations and all other systems, signals, or goads that may apply, as functionally grouped according to biological primals, adaptive integrations, and higher valuative representations as discussed in the previous sections of this chapter. Collectively all these work with partial consciousness and partial autonomy within the brain's neurobiology and ultimately through one's body as a fundamentally significant energizing source of force, motion, and change within the surrounding physical world--what must be recognized as a human causation.

Much of human motivation, though not all, is rooted in or molded by an inherent quality of localization in relation to external circumstances, referred to in this work as 'radiality'. In this, we humans are fixed from birth to death at an intrinsically imposed center of primary responsibility within our brain and body and relate to everything outward from this center. Our awareness, concern, and experience generally diminish with distance, be it physical or psychological, from this center. We are urged by need, desire, and curiosity to expand outward, to satisfy hungers, to grow, learn, wander, explore, engage, or bond, often in active association with imagination. The related concept of self is also important in motivation as indicated in Chapter 4. Both radiality and self are strongest in the biological primals, and progressively less so through the adaptive and valuative capacities. The degrees of radiality and self, particularly in these capacities, are plastic and variable across individuals.

2.3 The Significance of the Valuative Representations

Valuative representation is our most internal functional capacity. It is the home of reflective imagination, highest perceptual evaluation, reflective cognition, and global valuation. It is the central home of our consciousness and free will. It works most to our immediate practical advantage when we use its higher representations as maps of external circumstances and its judgments as higher guides to life decisions and planning.

More broadly, this highest of our capacities offers us: (a) a quintessentially vital range of partially conscious freedom around and in optional association with our individual sources of cultivatable value, understanding, and aesthetic appreciation, (b) the ability to create perspectives and guidance for any purposes in any contexts, (c) the means to place, satisfy, or resolve self concerns of all levels within larger perspectives relatable to surrounding reality in many buffered ways, (d) the means to pass easily from self concerns to concerns of larger objective, universal, and transcendent considerations. These valuative representation capacities in individuals are the seats of ultimate judgment in all human affairs, individual and collective.

Conscious experience is not exclusively associated with this capacity, but flowers there more freely and expansively than anywhere else. The biological primals and adaptive capacities also have clearly recognizable existential and operative qualities, but in valuative representation the existential is more primary, the operative more derivative than in these others. Herein resides both its strength and its unique significance for humans. This capacity provides an inner theater wherein one can remove oneself in an inner reflective space away from time and the tyrannous imperatives of biological and adaptive living. In this theater we humans can find the dimensions of rational mapping and truth, value, aesthetics, and proportion. We can use these dimensions of judgment to create our highest achievements and highest understandings, and as higher guides to life decisions and planning. We can imagine conditions beyond those in which we live. We can try to create approximations to these.

The capacities and qualities of valuative representation define our freedom of creative reflection and imagination, and our ultimate uniqueness and place in the whole of things as we can best come to see them. Indeed, this theater of free reflective valuation may be seen as the quintessence of humanity.

2.4 The Justification and Obscurities of Physical-Experiential Organicity

This chapter has presented an integrative functional view of the inner sensibilities which more fully and satisfactorily covers the full range of human experience and behavior than can be gleaned from considerations of either its neurobiological or experiential faces alone.

The structured view of physical-experiential functional organicity given here seems clear to universal common experience and introspection, and further justified by the extraordinary scope of harmonious integrative understanding which follows in its wake. We all feel as one with our conscious awareness and feel we autonomously direct the particulars of our living behavior, though buffeted and constrained we may be by larger external and internal influences. The experiential dimensions of aesthetics, cognition, and value are endemic to human experience, in both individuals and groups, as can be further attested by the introspection of any individual. Individual life and experience easily can be seen and individually felt as reflective of these functional realms of biological primals, adaptive integration, perceptual aesthetics, apprehension-cognition, and value-good and bad. The remaining chapters will show that explicit recognition of these capacities leads to a clean integrated view of the human nature, condition, and experience sufficient for the observed richness, diversity, and apparent contradictoriness of these.

It is not surprising then that the history, activities, and institutions of the peoples of the world also embody these same capacities as can be readily recognized with a moment's reflection. The biological primals are the focus of the rudimentary functional structures of all societies and civilizations. The adaptive integrations are the means by which we deal with our condition and wants, and the home of many skills and professions.

The embodiments in our cultures of our individual capacities of generalized perceptual aesthetic experience, reason and knowing, and value, goodness, and evil manifest communally as the arts and humanities, scholarship and universities, religious and spiritual centers--what this work labels as 'the communal enlightenments'. These are essential requirements for the development of cultures, and the hallmarks of civilizations. Conversely, these communal institutions are stable fixtures of humanity and attractive to individuals precisely because their roots are prominent intrinsic capacities in every human.

Endnotes

1. Jackson1932
2. Tudge2000, Gould2002, De Wall2001
3. MacLean1970
4. see a brief summary of brain operational structure in chapter 21 of MacGregor1993
5. Maslow 1968

Chapter 3

The Structure of Human Value

It is remarkable that value has been largely neglected in many intellectual considerations of human nature throughout the twentieth-century. This stems most directly from the explicit nihilism expounded by Nietzsche and the absorption of value within aesthetics by literary figures[1] and many philosophers. Longer roots of intellectual antagonism to value are found in the contrast between the necessary skepticism of rigorous science with the faith and looser credulity found in religion. Neuroscience has recognized value most notably in the rudimentary biological emotions and drives of laboratory animals. More recently empathy and hatred have been emphasized (see Chapter 14).

This work, in contrast sees value as a richly diverse and distinct dimension of the inner sensibilities, well beyond these pairings of it with one or another other qualities. This chapter offers a comprehensive theoretical view of the nature, ground structure, and qualities of value in the brain's neurobiology, and a systematic categorization of the most easily recognizable values. Values are grounded in this chapter in the inner sensibilities so as to indicate in chapter four their relations to the brain's neurobiology. Later chapters consider the operation of value in the integrated functioning of the inner sensibilities as a whole.

This work has found it desirable and, indeed, necessary to include value as a prominent capacity in the inner sensibilities because it appears as an obvious and rampant central ingredient in the daily living of any human individual including oneself, and equally so pervasive throughout the fabric of all collective human life. Further, value seems a widely rich and diverse family with members ranging from direct blunt animal and survival needs through practical desiderata to subtle mentations and ethereal conscious experiences. This work will take value to consist of many manifestations in both neurobiological and experiential forms, many intrinsic, many others learned and plastic, and yet others sometimes obscure or inconstant. We will take it to manifest pervasively both in itself and in association with perhaps all the sensibilities.

Value seems a distinct independent dimension of human sensibility in itself quite independent of both the cognitive and aesthetic dimensions[2]. There is an aesthetic quality of some existential values, yet value also carries its own dimension which is more often the central dominant concern. This means, for example, that value is ultimately impossible to define satisfactorily except in the terms of its own dimension. It can be indicated with words such as 'that which gives significance' or 'that which characterizes the good and the bad'. But all such indications are ultimately merely circular collections of words which at best amount to a pale reflection of the operative reality in itself. To understand value it is better to directly experience its operations in an appropriate personal situation which requires a difficult choice.

Human value is seen here as intrinsic to all humans (and indeed to many lower life forms as well) and to project into external reality by human communication and action.

Human value shows a multitude of forms which may all relate in some way to the overall global well-being of one's self and the whole of things, inclusive of all one's sensibilities, external interrelations, circumstances and contexts, apprehensions, inner associations and reflections, and anticipations. Value is in part radial in individual humans (self-centric, expansive, universal, transcendent) as indicated in Chapters 2 and 9. In this, an individual experiences values regarding one's self, intimates, local or large-scale groups, and universal and transcendent concerns, in any conceivable mix. Value is taken in this work to manifest integratively across the central capacities and faces of the inner sensibilities. All of value is ultimately rooted in and reflective of actions and interactions of human life, body physiology, neurobiology, and conscious experience. Salient primal components include survival, appetites, vitality, joy, sorrow, love, free functioning, satisfaction, empathy, understanding, justice, beauty, mercy, and many more. The ultimate quintessential experiences of value include the rare but long-remembered peak experiences of happiness and transcendence available to humans. Values are central to human motivation, and Maslow has stressed an existential distinction in human psychological motivation between deficiency-driven and growth-driven behavior[3], which this work sees as a wide integration across all the inner sensibilities from biological foundations to being values, peak experiences, and higher sensibilities.

Figure 3.1 begins a map of progressively abstracted labels used throughout this book to structure its picture of the relationship of the observed values of humans to underlying locations in the inner sensibilities and the brain's neurobiology. The existence of such an overall mapping is a central theme of this book. Thus, for example, column C in Fig 3.1 identifies five root types of values which can be traced back through the inner sensibilities in column B which are in turn related to neurobiological roots in chapter four. Column A anticipates this by indicating the implicated physiological processes of the brain which are described further in chapters four and five. These connections in Fig. 3.1 are useful in suggesting, for example, that values associated with the biological primals are largely intrinsic, whereas those of the higher sensibilities are largely plastic.

Column D fixes a consolidation of these five types which will be used throughout the book. Column E breaks these down into seven classes labeled to represent collectively all human values and indicated in Table 3.1 by a large collection of easily recognizable human values for each label. This mapping is continued in the next chapter in Table 4.1(b) and surrounding text which more explicitly relates values to the brain's neurobiololgy [through classes and sensibilities of columns D and A in Fig. 3.1 and Table 4.1(b)]. All this, again, is a rough but respectable and useful map of the neurobiological undergirdings of the values shown in Table 4.1(a) consistent with current neuroscientific knowledge and thinking. The map, though tentative and rough, is especially useful because value as such has not been treated systematically in neuroscience. It is further complemented with rough, overall tables of human values recognized by philosophy and religion in

Chapter 11. The obscure relationship between the experiential and biological faces of value, as that of consciousness itself, is discussed in Chapter 6.

3.1 Five Root-Types of Value

The center column (C) of Fig. 3.1 shows five root-types of value as rooted separately in the three main sensibility components. These latter and their underlying neurobiological dimensions are indicated in columns A and B. This view sees two root-types of intrinsic value rooted in genetically prescribed physiological and neurobiological processes of the biological primals--personal primal and sociobiological value which are distinguished from each other by the intrinsic interpersonal involvement of the latter. Some personal primal values have additional linkages with structures of adaptive integration. Some of these are indicated among the so-called 'being values' introduced some years ago by Abraham Maslow[3] (shown in Table 3.1).

The intrinsic forms and processes of the valuative representation capacities in themselves (aesthetics, reason, value in itself)--as separate from any content upon which they operate--are taken as a third intrinsic root-type of value, valuative-representation-essences. These are in their fuller reaches inherently transpersonal and thereby transcendent. The essence of the value dimension in itself contains a particularly significant realm of 'peak experiences' of individual joy or well-being and fulfillment identified by Maslow (also shown in Table 3.1).

The fourth root-type of value consists of personal-reality values which are involved in the living of one's life in interaction with external reality, part of the adaptive integration capacity. This root-type is fundamentally a mixture of intrinsic and plastic elements. It is rooted in the intrinsic processes of the adaptive integration (will, intention, adaptation, ..), and operates intrinsically in interrelation with both the biological primals and the valuative representation capacities. It manifests as the main definer of the most practical, external, worldly-related domain of value--the sociopersonal--in a mix with contributions from all root types.

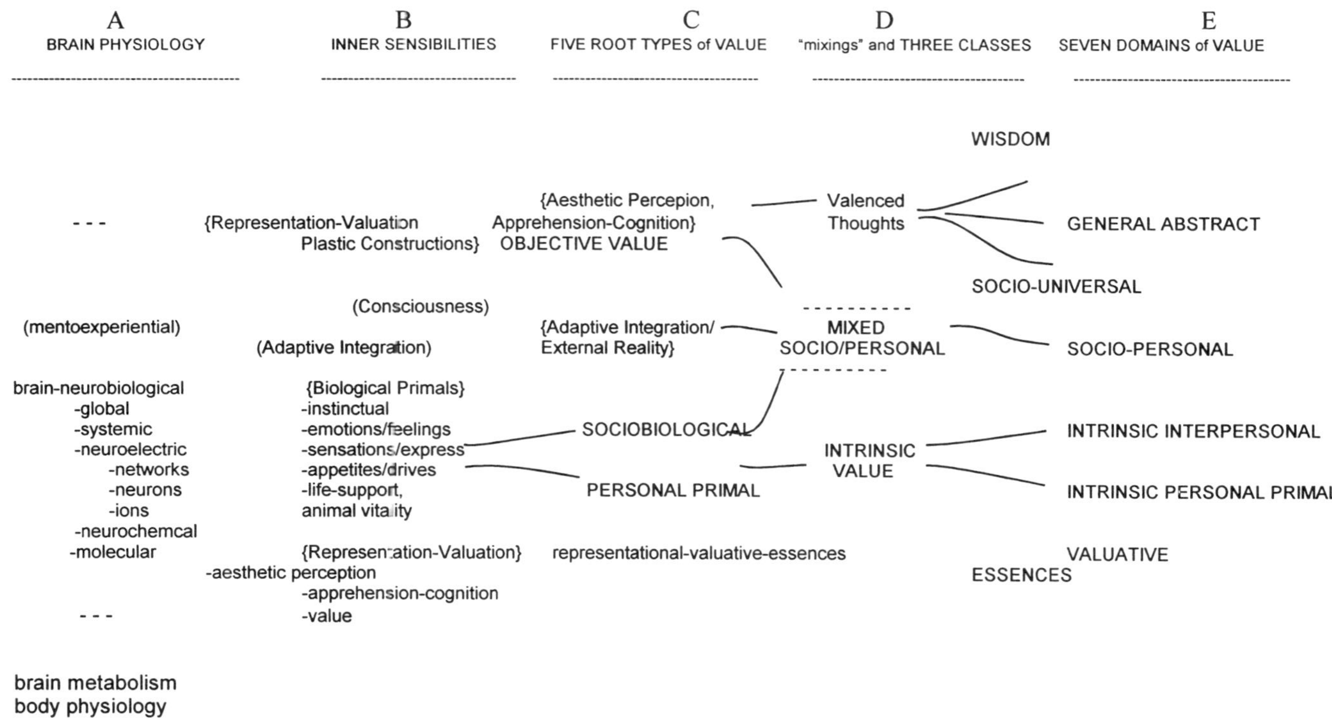

Fig. 3.1 The structure of human value.

The operations of the primary valuative representation capacities on particular content produce a fifth root-type of value, objective values. Objective-values reflect the internal pursuits of 'what is this?', 'what is true?', 'what is good?' These questions represent the operations of the three valuative dimensions--aesthetic, cognitive, and value itself. These include both present circumstance and recollections and reflections--learned, malleable, plastic, tentative, and open-ended. This root-type is the central source of the third main class of values--valenced thoughts. These are produced by the dual-process associations of cognitive and valuative elements which is a functional 'value-weighting' or 'value-valencing' of particular cognitive representations by particular elements of value. In this the word 'valenced', is used as equivalent and interchangeable with 'charged' in the sense of being endowed with an energizing quality, parallel with the use of the word in chemistry to describe the non-neutral electrical state of ions (molecules who have lost their electrical neutrality by the loss or gain of electrons).

Valenced thoughts include finely discriminated human values, being associated particularly with the discriminative processes of analytic thought. In the formation and valuations of cognitive elements in themselves, the processes and judgments of cognition (rationality, truth, coherence, correspondence, contradiction) are dominant, sovereign. However, in the realm of value, including valenced thought, value is the major player, usually the dominant, sovereign. This includes the value of intrinsic truth, but is not restricted to it alone. Wisdom is fundamentally related to human values, consisting largely of valenced thoughts. That is, these thoughts are energized with value. Like electrical charges, these valences can be positive or negative.

There are several intrinsic qualities of valenced thoughts which result from the intrinsic partial and plastic nature of the cognitive processes. First, all valenced thoughts are tentative, contingent, and subject to correction--they are necessarily components of an open-ended pragmatic system.

Second, almost all valenced thoughts are half-truths in themselves: As any given human situation is uniquely endowed and ensconced within particular circumstances, it requires the confluence and balancing of any number of valenced thoughts with proportions appropriate to the situation at hand.

Third, individual valenced thoughts and complexes must seek their own levels within continually recycled validation testing within the unfolding life experiences of individuals, similar to the experimental validation testing of scientific laws and theories[4]. This implies a continually recycling pressure towards self-corrective refinement or growth. In summary, the three functional capacities of the inner sensibilities produce five root-types of value (two from each of biological primals and valuative representation, and one from adaptive integration), in three classes--intrinsic, mixed sociopersonal, and plastic, as shown in columns B, C, and D of Fig. 3.1.

3.2 Seven Domains of Value

Column E of Fig. 3.1 shows seven domains of value drawn from the five root-types. Valenced thought is the central root-type for three resulting domains--wisdom, abstract, and socio-universal. The other four root-types each serve as the central source for a single domain--valuative-representation-essences, intrinsic personal primal, intrinsic interpersonal, sociopersonal. In practice, there is the potential of some cross-mingling of root-type influences across all domains. The definition of three domains of valenced thought is somewhat arbitrary, but cogent.

This structuring, shown in Table 3.1, is a second main contribution of this chapter. Representative content for each of these domains is shown in Table 3.1. This connects the labels introduced in Fig. 3.1 to recognizable values of human life, and thereby partially fulfills a central goal of this work to associate the brain's functional organization with recognizable concerns of daily life. This table, although representative rather than exhaustive, speaks for itself and should be perused at this point. This collected description of recognizable human values is completed in this work with the tables of chapter eleven which show values recognized in philosophy and in the worlds religions, and by Fig. 14.1 which outlines a dozen central streams of good and evil in the human inner sensibilities. The connection to neurobiology is continued in Chapters 4 through 7.

Table 3.1 Domains of value — Part 1.

Type, class, nature	Representative content
valenced thoughts	• WISDOM-{... senses of the underlying currents of human existence on earth and in social interrelation, one's nature, place, achievement, degree of engagement-alignment with larger currents...}
valuative representation plastic, learned ,partial, incomplete, tentative, pragmatic, open-ended. principles & complexes. combines cognition and value	• GENERAL ABSTRACT-{communal enlightenments: knowledge, education, philosophy, science, arts, religion, ... }
	• SOCIO-UNIVERSAL-{peace, security, order, civility, cooperation, justice, enlightenment, achievement, law, progress, non-violence}
sociopersonal values adaptive integrations mixed: combines valenced thoughts and intrinsic values	• SOCIOPERSONAL-{security, place, decency, civility,... acceptance, tolerance, respect, benevolence, virtue, compassion,...non-abuse/ autonomy, adaptation, homeostasis, achievement, .../ self-aggrandizement, ego-attachment, avarice / rivalry, greed, bigotry, envy, injustice, slander, domination, abuse, ...}
sociobiological values genetically prescribed or inherent	• INTRINSIC INTERPERSONAL-{attraction, like, trust, love, root kindredness, compassion / aversion, distrust, dislike, repulsion, ..disgust, animosity, envy, hatred, malice/ ..interest, curiosity... radial expansion ... karmic aftermath (good and ill)/ finding: support, recognition, acceptance, tolerance, rejection}

Table 3.1 Domains of value —- Part 2.

personal biological genetically prescribed biological primals	• INTRINSIC PERSONAL PRIMAL-{animal vitality, fun, humor, appetites, bodily pleasure, satisfaction, contentment, happiness, joy, curiosity,..fulfillment/ melancholy, moods, feelings, excitement / pain, want, sorrow, grief, anxiety, fear, anger...} being-values--[self-sufficiency, honesty, playfulness, effortlessness, uniqueness, grace, goodness, simplicity, richness, aliveness, justice, completion, perfection, wholeness]
intrinsic valuative representations genetically prescribed dimensions: cognition, aesthetics; operational	• VALUATIVE REPRESENTATIONAL ESSENCES-{goodness (value), truth (apprehension-cognition), beauty (aesthetic perception)} peak experiences of happiness & fulfillment--{athletic, somatic, insight, creativity, adventure, aesthetic experience, parenting, bonding and friendship, love, mystic experience--[peace beyond understanding, ultimate meaning, preciousness of all life, infinite sweetness, ultimate commonality of all life before death, all-inclusive benevolence, love]}

Peak Experiences, is listed as one of the inner valuative-representation-essences, because of its unique power to impress and influence overall life orientation. Valenced thoughts are tentative, pragmatic, and open-ended in reflection of their essential association with the cognitive-reasoning capacity. It is important to see that the dimension of value includes both positive and negative components.

3.3 The Peak Experiences

The values listed as 'peak experiences' in the last lines of Table 3.1 are deep, sometimes life-changing, and sometimes ultimate personal experiences[3]. Such experiences are significant influential values for most of us, and, for some, the deepest sources of meaning and direction in life. All these experiences are readily found in one's own personal experience or widely documented in collected human experienced or literature. Nonetheless, some of these experiences may be only rarely, partially, marginally, subliminally, or not at all a part of the experience of some. The deepest reaches of these experiences carry expansive transcendent senses of external inclusiveness and transpersonality. These latter features are perhaps most fully and pervasively embodied in the ultimate mystical experience. The inherent obscurity, variability, and rarity at these deep foundations of individual experience is a significant feature of human life.

3.4 Comment

This structuring of values underscores that individual humans intrinsically operate with a radial mix of personal, intimate, local or large-scale group, and universal and transcendent concerns. The interrelation and balancing of values from various domains is a vitally important, open-ended, and never-ending dimension of human existence -- individual and collective.

Endnotes

1. Nietzsche1886, Joyce1916
2. The English philosopher George Moore also advocated the non-definability of the Good, see pages 372-381, vol 5 in Edwards1967.
3. Maslow 1968
4. Popper1959

Chapter 4

The Inner Sensibilities and Neurobiology

This chapter introduces basic neurobiological material as background for the next seven chapters. The first half of the chapter considers: higher integrative plasticity; evolution and brain organization; functional topography of the brain; motivation and human causation; hierarchical stratification of brain variables; and consciousness and neurobiology. The second half outlines the neurobiological relations of the inner sensibilities as we have introduced them in the two previous chapters.

4.1 Higher Integrative Plasticity in Humans

The human genes (contained in our DNA molecules) construct and govern both the anatomical structures and physiological processes of all the body including all those of the brain and nervous system, which are the physical foundations of our consciousness and inner sensibilities[1]. The genes of any given individual include substantial material common to the species (all but one part in a thousand) and some particulars common only to one's close family, and some only to one's self, in roughly equal proportions from each of one's two parents. The fine-grained structure of these genes (and correspondingly the structures and process they govern) may be seen as the products of a continuing long-term population process of selective change, wherein alterations of gene structure brought on by any process (often random mutations) are passed on to progeny according to the principles of natural selection[2].

This process of selective evolutionary alteration is a primordial natural plasticity of the anatomy and processes of virtually all life, including the human brain--a fundamental plasticity of the very hardware of functioning of the body and the inner sensibilities. This is a slow, long-term plasticity, operating across generations, which reconciles mostly chance alterations of function with biologically-relevant characteristics of our ambient earthly surroundings.

Humans in particular, also exhibit, to a pervasive extent well beyond that of any other species, a higher experience-dependent plasticity of inner constructed patterns, many of which help govern one's interpretations and interactions with the external world. This work labels this the 'higher integrative plasticity' or 'neural plasticity' of the inner sensibilities, and its elements 'inner constructions'. This is a second kind of functional plasticity, one which brings with it an entirely different realm of inner functionalism and inner control. It operates on the short term--within the hours and years of an individual life and is extended outward by interpersonal behavior and communication. It is unique within the particulars of accumulative experience and temperament of individuals. Significant training and experiences of the very early years of life are typically deeply and tenaciously ingrained in the constructions of this plasticity and many of these seem to form cornerstones of one's overall inner scheme of the ways of the world and subsequent integrations relative to these. The inner patterns of higher integrative plasticity continually incur subsequent self-correcting modification and realignment within the ongoing operations of the inner sensibilities as a whole relative to surrounding influences. We can use this plasticity to construct independent inner representations, and then plastically combine or modify these at will. These processes can partially ameliorate early negative conditioning and especially buffer these within larger higher order adaptive inner constructions. This higher plasticity in individuals is extremely more rich, finely filigreed, and idiosyncratic than is genetic prescription of, say, instincts or archetypical templates.

Significant experience-dependent plasticity occurs widely and variably across the animal kingdom, including that associated with the higher regions and higher functions of the brain. However, humans have this latter to a much higher extent than even the great apes, our closest phylogenetic predecessors. It is most likely grounded in the structuro-functional malleability of vast synaptic interconnection systems of the mostly cortical neural networks of the brain. The terms "Neural Darwinism' and 'Neuronal group selection' have been introduced by Edelman[3] to describe the continued maintenance and use of a plastically generated neuroelectric circuit pattern within neurobiological networks according to its usefulness in its holder's experience and behavior, as discussed in chapter five.

Our DNA differs from that of great apes (chimpanzees, orangutans, baboons, and gorillas), for example, by only about two percent despite our much larger masses of cortical tissue and corresponding exceedingly larger magnitudes of particular content in motor skills and especially representational discriminations. This may underscore the watershed significance of our dependence on this higher trans-genetic plasticity. This two percent difference may well house the genetic foundations of our plastic functions--our central operative distinction; the learned plastic embeddings themselves, however bountiful, are in themselves beyond genetic prescription and require no space in DNA. This integrative plasticity and its coupling with conscious partial autonomy is the hallmark quality of our species. Its vast inner constructions are the ground stuff of our extensive adaptivity and our overall direction of living. As most human qualities, this plasticity is inherently mixed, partial, incomplete, variable, and constrained within limits.

4.1.1 *Open-endedness and adaptivity*

It should be recognized at the outset that the full range of this integrative plasticity provides a quintessential open-endedness to human inner control and direction within our earthly existence. This open-endedness is vastly more amplified in humans than in any other creatures because of its intimate association with our inner abilities to construct, plan, and evaluate. Open-endedness is central to all human individuals and human collectives in which latter its potential is again significantly expanded.

Adaptation of human behavior to surroundings and situations is a vital rudimentary quality of both genetic and integrative plasticity. 'Conditioning' and 'trial-and-error' learning are widespread in humans and occur to lesser degrees in some higher animals. In this, we can see experience-dependent neural plasticity as a watershed evolutionary step which supersedes instinct as a foundation for the control of more complex behavior. Sociobehavioral processes regarding the needs of biological life on earth are largely governed in higher animals (and in our deeper biology) by means of instinct which generally uses 'fixed action patterns' and various archetypical perceptual templates to identify 'trigger stimuli'.

4.1.2 *Self-direction, autonomy, consciousness, and evolution*

Autonomy and consciousness are both partial and obscure qualities at the very core of the inner sensibilities in humans, and even more obscure in lower forms. It is easy to imagine a joint flowering of these in conjunction with integrative plasticity, open-endedness, and especially self-direction as a major evolutionary selection. In this, both consciousness and partial autonomy are intimately involved in the construction and use of extensive plastic patterns we use to direct our behavioral learning. This view is developed in chapters seven through ten. From this evolutionary point of view, partial autonomy and partial consciousness are the alternative to instinct as means of constructing and directing complex behavior.

Our biological primals are especially similar to those of mammals. Our mixed-plastic adaptive behaviors are rudimentally similar to those of many mammals, but much more richly developed. Higher valuative representation capacities seem highly developed only in humans, with only the most rudimentary suggestions in the great apes.

Neurobiology dominates more towards the more rudimentary levels, and the mentoexperiential more towards the higher. Consciousness is especially related to the adaptive and valuative capacities and even more especially to their plasticity. It largely controls attention which, in turn, strongly facilitates higher representational learning. Existentially, we identify with our conscious awareness and autonomy.

4.2 Evolution and Brain Organization

In the broadest terms, one may see the evolutionary production of new species in nature as a primary form of generative renovative plasticity (alterability, adaptability, changeability) of the genetic hardware which produces and guides the structuro-functional organization of living forms, and which may entail any or close to all modules of structure or function. This is an important quality of life which helps life as a whole adapt to local circumstances and to progress to higher capabilities. These adaptations and progressions are relatively slow, operating over generations. They consist of rigidly-fixed genetically-prescribed patterns of structure and function. The fixed species-dependent behavioral patterns of 'instinct' are salient examples of these in the nervous system. Instincts are generally closely associated with and abet particular idiosyncrasies of a species' body form. The recognition of evolutionary change as serving adaptability and progressive enhancement of life forms is important for our purposes because it contrasts our human neural plasticity with the predominant genetic prescribing of virtually all other life forms. Only in mammals does much extra-genetic adaptability occur. No other life form enjoys the extensive dependence on neural plasticity as does the human genome and life strategy. The DNA (genome) of individual humans differ by less than one part in a thousand. Our DNA differs from that of the closest great apes, chimpanzees and gorillas, by about one part in fifty. Chimpanzees can learn a vocabulary of about four hundred words, roughly that of six-year old human. They make tools, adapt, and seem to have some imagination and inner reflection, but, one might guess, to smaller extents and fullness than do humans.

Thus, evolution may be seen as a plasticity of the genes (DNA) which produces adaptability and progressive growth of life forms by means of relatively rigid prescriptions of structuro-functional modules, including, in humans especially, the production of a second type of freer neural plasticity. We take it that neural plasticity, adaptability, self-direction, consciousness, partial autonomy, and open-endedness, although all existing to some lesser extents in earlier forms, underwent an explosive collective evolutionary flowering in humans which defines our essential existential and functional nature. All of this is particularly associated with the construction of a vast system of neural 'inner constructions' for the internal representation of things and their behavior, including, but by no means restricted to, the practicalities of life and survivability. This work has in fact recognized both adaptive integrations and freer valuative representations as major components of the inner sensibilities to emphasize the freer nature and special transcendent universal qualities of the latter.

Neural plasticity operates uniquely within individuals according to the particular experiences and conditions of one's own lifetime. It does not in itself share with others or pass on to progeny except through interpersonal behavior and communication. Evolutionary genetic plasticity is highly wasteful in that candidate generative renovations and new species are produced by random and trauma-generated mutations the vast majority of which are diminishments which impede successful reproduction and rapidly die out. Neural plasticity is vulnerable to vagaries of early conditioning experiences and training much of which can be limited, misleading, wrong, or debilitating. On balance, however, neural plasticity introduces vast amounts of incalculably useful immediately life-relevant learning, the ability to create constructive responses to new situations, and the life-long ability to modify and self-correct established action patterns. It also produces a rich personal internal reflective world of imagination and creativity which provides the centerstage of individual human existence. These qualities of neural plasticity open up the human potential to vistas far surpassing those based on genetic-prescription only.

The mechanics of evolution may be visualized as an exponentially-expanding bush in which each candidate species is a potential source node of many branches, each branch another candidate species and potential node of yet more candidate species (each another branch). Any species is thus the potential source node and common ancestor of such a bush of species. Evolution may proceed continuously, for a period, locally within a single or few species--as, for example, when a species is moved to a new environment or a smallish change of climate occurs in a local environment. Alternatively, and perhaps predominately, evolution may also occur in a pattern of 'punctuated equilibrium' wherein scattered bursts of very large numbers of mutations (new candidate species, new branches) are followed by the rapid winnowing away of unviable species to an equilibrium of better suited species which continue to reproduce for many generations until some other big burst of mutations occurs. The origins of such bursts are not always clear but include environmental changes or other extraneous intrusions such as surges of radiation.

In this bush picture, any branch reflects usually a relatively small mutation from its immediate predecessor, and may entail as many such small changes from any given source point as there are intervening branches. Scientific classification of life forms (taxonomy) is still largely guided by overall somatic morphology and the concepts of 'single common ancestor' for significant groupings and, recently, 'clades' for those groupings which also contain all subsequent heirs of the common ancestor[2]. Cladistics is the rigorous contemporary approach to scientific classification which integrates molecular DNA findings with traditional classifications. Mammals and reptiles are both seen as reflective of single common ancestors, but mammals are a clade and reptiles are not. There are more variations among species in the nervous systems of vertebrates than in any other organ of the body, but nonetheless the central nervous systems of vertebrates have a common structural organization (which can be best seen in embryonic stages of advanced vertebrates) whose overall morphology is readily recognizable in reptiles and mammals including humans[4]. Hughlings Jackson has suggested the principle of 'levels of construction' which asserts, beyond this common structural (genetic) foundation, that the electro-chemical organization patterns of at least the vertebrate nervous system determine a certain self-sustaining tenacity of some larger-order functional constructions with respect to subsequent genetic variations in particulars[5]. This principle anticipates contemporary thinking that in the nervous system function is not only guided by structure but sometimes predominates over or guides it in understandable ways. With or without Jackson's principle, this common overall structural organization is useful as a rough indication of the localization of function in vertebrate and especially the human brain, and is consistent with many laboratory and clinical studies.

4.2.1 *The tyranny of survivability*

The only value universally recognized in evolutionary science is survival value. The universal monopoly of this criterion to the total disregard of any other qualities is surely an overstatement of a fundamental truth. This is something like recognizing only the necessity to eat in evaluating the selection, care, and enjoyment associated with eating, or only the practical necessities of life rather than its fuller ranges of existential, imaginative, and creative dimensions. The tripartite characterization of brain function by the inner sensibilities as biological primals, adaptive integration, and valuative representation (generalized aesthetic perception, apprehension-cognition, and value-goodness) is a better match to the observed experience, concerns, activities, works, and cultural achievements of advanced civilizations than many scientific reductive views. This is particularly apparent in the richness of the aesthetic, extra-cognitive, and valuative constructions of individuals and cultures as seen, for example, in the communal enlightenments (Chapter 12). Just because something is necessary for existence or survival does not mean that it is the only thing of significance. Our open-endedness of imagination and creative construction in particular provide for levels of achievement and meaning within humans far beyond our basic survivability in itself. Indeed, in a slightly different perspective, humans frequently attach high levels of significance to the survival of non-humans, and individuals may sacrifice their own survival for others or for universal or transcendent ideals.

4.3 Functional Topography of the Brain

The brain may be broadly seen to consist of an integrating shallow plane (cerebral cortex) crumpled over a bulbous cap (limbic system) surrounding the top of an upward extension (brain stem) of the spinal cord. The spinal cord, together with peripheral nerves and the autonomic nervous system, performs many essential sensory, sensorimotor, rudimentary integrative, motor, and effector operations. The brain stem houses much of the central control of the biological primals--including higher order sensorimotor reflexes, arousal, appetitive behavior, instinctual regulations, biorhythmic control, and portions of the autonomic nervous system which itself controls most of the body's rudimentary life-support functions and rudimentary biological responses and regulations. The midbrain tegmentum and thalamus are involved with correlative processing of these rudimentary functions. Instinctual patterns are housed in the hippocampus, midbrain, and reticular formation. The limbic system is associated with affective features of experience such as reward and aversion, visceral states, and emotions, and, in its higher paleocortical regions (mainly hippocampus), with rudimentary representation, cognition, experiential memory, and dreaming. In particular, Papez outlined the 'Papez loop' through this limbic region as the substrate for emotions[4], and Penfield identified a chronologically disposed bed of one's experiential history in the perihippocampal region[6]. The anterior (forward) cortical regions are associated more with overall behavior and governance. The posterior (rearward) cortical regions are associated primarily with representations more generally, including particularly sensorily-derived representations of the external world and its qualities. Both regions have areas closely related to externals through motor and effectors (anterior) and of sensors (posterior). They also each have deeper internal regions, functionally closer to each other, which deal with more abstract generalities related to plans and expressions (anterior) and representations (posterior). Structures of language, for example, form an interconnected substratum mostly across intermediate and peripheral areas of both anterior and posterior regions[4].

Jackson's concept of 'levels of construction' asserts that in the evolution of the nervous system, certain stable levels of overall functional organization have been attained, the functional integrities of which have been maintained in subsequent evolutionary development. MacLean adapted Jackson's principle in his triune brain theory which recognizes a neocortical level of control in humans, superposed over a lower level of emotional limbic control dominant in lower mammals, in turn superposed over a lower level of instinctual control dominant in reptiles[7]. Vertebrate taxonomy and morphological and much physiological study are consistent with this as a rough outline. (See the last paragraph of the second preceding section just above.) Luria based a richly described view of overall human functional organization in terms of three functional systems[8]: a unit in anterior cortex for overall regulation and programming, a unit in posterior cortex for receiving, storing, and processing information, and a unit in the limbic system and brain stem for regulating tone. Each cortical unit is subdivided into three levels of depth. Right brain is associated with global diffuse qualities and left brain with discriminating local qualities in both cortical units, progressively so in the deeper regions. More recent views of global organization have emphasized the diffusivity of higher functions throughout wide ranges of particularly cortical regions. Experimental neuroscience and biopsychology give much explicit detail on the localization of functions throughout the anatomy of the nervous system, as are summarized in many recent textbooks.

4.4 Operational Organization of the Brain

Deeper thinking about the brain wants not simply a functional topography but an operational functional understanding. Motivation, energization, and the brain's multi-leveled hierarchy of physiological processes, which we will describe more fully just below, are the realms to which all views of brain and conscious organization must ultimately be resolved[9].

4.4.1 *Motivation and human causation*

We take a generalization of motivation as equivalent to a humanly directed causation and see this as a fundamental ingredient of the human inner sensibilities. It is intimately associated with our partial autonomy, consciousness, and the mind-body question, and shares the obscurity of these. Our partial autonomy especially, with its sense of a grounded, constrained but partially free will, is one major face of this foundational human causality. This causality is a type of energization, centrally significant in the affairs of individuals and humanity, and, through us, increasingly significant in the earth as a whole. Although significant to humans and to much of earth, it is of less apparent significance to the universe as a whole, and its relation to other forces of nature--gravity, electromagnetism, weak, and strong forces --be it derivative, dependent, localized, or parallel and independent--is obscure. (See Chapter 6).

This larger view of human energization and human causality leads us to a significantly broadened and inclusive view of human motivation. Terms like motivation, drive, homeostasis, regulation, adaptation, intention, come together with volition, preference, happiness, value, objective inner reflection, goodness, truth, beauty, justice, and many others as manifestations of various faces and levels of human causality. Drive and the like correspond largely with the animal-like biological primals. Intention is associated largely with practical concerns of personal adaptive integrations. Senses of objectivity, preference, volition, value, universals, and transcendentals and Maslow's growth motivation are associated primarily with the valuative representations[10].

4.4.2 *The hierarchical stratification of brain variables*

The bottom-up approach to overall brain operation from the rich hierarchy of brain physiology, however, is fundamentally impeded by the brain's sheer complexity. It is also seriously hindered by the thorny problem of apparent partially conscious autonomous top-down direction, which seems to demand recognition as a major, although highly obscure, player in this very operational organization. Alternatively, middle-level approaches might make use of human causation and the motivational systems discussed just above. These, or middle-level constructions like Freud's id, superego, and ego, or our tripartite inner sensibilities, or any other metaphors may indeed be useful for certain psychological or behavioral interpretations, but are ultimately incomplete, needing to impose some overall governing principles including some view of consciousness and to establish physiological undergirdings. On the other hand any top-down view of conscious control is in itself woefully incomplete without recognition of all the multitudinous things the brain does without consciousness, and also wants some indication of the roots of consciousness itself in relation to its brain.

Any given physiological process of the brain is inherently embedded in an integrated multi-leveled physiological hierarchy, influenced by and often directed towards external reality, and possibly associated in some way with consciousness and thereby underwritten in obscurity. Table 4.1(a) defines five fundamental dimensions relative to brain and conscious functioning and elaborates especially the inner physiological and conscious levels of brain operation[9]. This stratification of variables systematically identifies the interrelationally structured undergirdings of brain operations which we describe more fully in the next chapter. It also identifies the ground structure of the obscurities of brain-mind relations discussed in Chapter 6. The stratification is also particularly useful in contextualizing larger questions regarding functional organization and in studying its operations as we will now illustrate here.

An essential feature of this arrangement is that each stratum seems to have its own principles and processes of operation, logic, and cohesion, and at the same time to stand in some significant influential relationships (either dependent or supportive) with the strata below and above it. Several large-scale questions emerge from this arrangement: Almost any significant brain process is manifested simultaneously in coordinated physiological processes occurring in each or most of the seven neurobiological levels shown, with implied or possible involvement of some of the extra-brain levels. A fundamental question regarding the functional operation of the brain asks as to the relative degrees of upward and downward influences, and of autonomy, among these distinct levels of stratified control. Table 4.1(a) encourages us to question the nature of dynamics and function associated with overall functional organization across such levels as the molecular, neurochemical, ions, neuroelectric currents, neurons, neural networks, and neural systems.

A fundamental question asks as to the interpretation of any correspondence between processes in distinct levels active in a given event. One of the deeper questions suggested by Table 4.1(a) asks of the place of autonomy and consciousness within this spectrum, and of the root origins of these in themselves. These last two questions relate to the apparent paradox of partial autonomy operating on and within a neurobiological system presumably subject to determinative physical law. This stratification points to multi-level two-way causal interrelation, in a fashion directly relatable to physical theory and physical law. Theoretical neuroscience must be ultimately rooted in neurobiological processes, as our means of understanding the mind's relationship to physical law--to ignore or lose sight of this is to lose ultimate scientific foundationality. (See Chapter 6)

Table 4.1(a). The hierarchical stratification of brain variables.

- total external reality

- inner sensibilities: conscious and neurobiological
    - global
    - systemic (composite networks, sensory,motor, language)
    - neuroelectric
        - neural networks
        - neurons
        - membranes, ions
    - neurochemical (synaptic transmitters, neuroregulators)
    - molecular

- life structures and processes
  (metabolism, respiration, endocrine, neuroglia, ...)
- physical reality
  (matter, energy, particles, waves, strings, physical law)
- (ultimate mysteries)

4.5 Consciousness and Neurobiology

This section introduces some rudimentary attributes of consciousness as preliminary to the fuller considerations given in the next three chapters. The stratification of variables introduced earlier in this chapter clearly indicates fundamental questions of energization and causation implicit in the relationships between consciousness and 'top-down' organization in living activities, on the one hand, with physical lawfulness and its 'bottom-up' influences on the other, and the foundations and scope of autonomy and free will in this. Interactions between consciousness and neurobiological processes are two-way. Consciousness tends to participate in the direction of movement patterns, life plans, thinking, and many other activities, rudimentary and complex. On the other hand, consciousness is dependent upon both body biology and brain sensibility processes for its very functional integrity, and is generally a rather passive receiver of perhaps the strong majority of its normal content, including sensory and bodily sensations, emotional feelings, and content of aesthetics, cognition, and value.

Consciousness seems to permeate most anatomical and functional capacities including all three of the main functional capacities: biological primals, personal integrations, and valuative representations. However, all three groups of capacities also have both significant subconscious and genetically-prescribed contents. Moreover, consciousness seems to be more weakly associated with the biological primals than with the other two capacities. In this, consciousness might be seen as especially related to the qualities of human overall regulation, direction, and internal reflection and even more especially to the plasticity of these. Consistent with this is the observation that attention is a sensibility largely (but not totally) controlled by consciousness. This is remarkable in itself and also because higher representational learning is strongly dependent on attention.

It is important to be clear that the inner sensibilities has its own operational directions and momentum beyond those of conscious control or experience. Many important events go on or occur unconsciously, ranging from rudimentary processes of biological primals to highly significant decisions, integrations, and creative constructions and valuations. Within this larger operative whole, consciousness exerts only partial control on sensibility processes. The inner sensibilities is more fully active even when consciousness is involved, and consciousness doesn't always dominate actions nor decisions.

In all of this it is well to remind oneself of the primacy of our personal conscious experiential sense of being and life as discussed in chapter nine. Conscious awareness is our necessary point of origin and existential placement throughout our lives. It is the site of our most precious senses of existence, autonomy, meaning, and life itself. It is our sole gateway to both the cultivation and expression of these, and our wider experience with surrounding reality whatever this latter may be, and whatever may be its secrets of existence and energization.

4.6 The Inner Sensibilities and Neurobiology

Our recognition of the existential and functional nature of the brain as the 'inner sensibilities' of biological primals, adaptive integration, and valuative representations is an intermediate level of specification between the fundamental physiology hierarchy and overall global control, to both of which we reach out in the subsequent chapters of this and the next part of the book.

Taken merely at its face value, this book's tripartite view may be compared, for example, with Freud's theory of the mind as composed of a struggling id (biological drives) held in by a repressive overseeing superego (social conventions and early conditioning), all mediated by an intervening ego (a pragmatic self)[11]. Freud's model is intrinsically operative, sees the functioning brain as wastefully struggling against itself, and is rather narrow in its view of human nature and potential. Alternatively, the inner sensibilities is also operative, but sees its various components as inherently harmoniously integratible with each other, and sees higher vision as inherent to human nature.

This work sees the concept of the self as an operative agent at the biological and adaptive levels, and as both operative agent and object at the higher levels of sensibility. Self manifests in association with biological needs and appetites, and especially survival at the biological level. Freud saw the self largely in terms of the biological drives and their participation amongst adaptive levels. At the adaptive level, self manifests in association with larger patterns of adaptation, especially including inner constructions involving social situations and interact-tions. Indeed at this level, the self is largely a social construction, often constructed by and of more use to others as much as by and to oneself. Lacan, for example, has stressed the adaptive, social nature of the self, noting particularly the absence of unity of self due to its multiplicity of separate factors or components according to the many various modes, realms, and circumstances of most individual lives[12]. Among the more free-ranging valuative representations of the higher inner sensibilities, the self is first an object to be harmonized and made integral, and an operative, to express one's larger concepts of life and place. Most cultures and religions stress the need for an overall integrity of self, including things like honesty and courage. Other spiritual teachings, while accepting integrity as a requisite operative base, have emphasized the personal self as a distractive and disruptive factor in higher universal concerns. This work sees the self as a hierarchical associational complex of plastic inner constructions (see Chapter 8) involving all three levels of the inner sensibilities, and existing ultimately as subordinate within these inner sensibilities themselves.

The inner sensibilities can also be compared to the cognitive, information-processing view of the computer metaphor for brain function described in the next chapter. This computer metaphor sees the brain as a computational cognitive device in all its conscious and subconscious representations and operations. In contrast, the inner sensibilities recognizes fuller ranges of conscious experience and full contact and dependence on underlying neurobiology, provides the natural fuller internal brain contexts for cognition, better represents the fuller purposes and larger nature of humans, and points the directions for the foundations of consciousness in its underlying physiological and physical undergirdings.

4.6.1 *Organic cooperation of the inner sensibilities*

We see consciousness as directing the highest control of inner constructions and behaviorally-related brain function. Additionally, many brain systems and subsystems have their own internal energizations, inner logics, forward dynamic momenta, and thereby their own motivational dimensions. Indeed, the inner sensibilities is a loose intracooperative tripartite functiono-structural entity whose three main capacities may each act independently, cooperatively, or in subordinate service, superordinate direction, or regulation of the others. There is a progression of direction from the most rudimentary animal-like operations and regulations of the biological primals, to the more autonomous adaptive directions of personal integration, to the free reflective valuations of the valuative representation capacity.

Each of the three main sensibility categories includes a rich multicapacitied supply of component operational processes and structures. All the following overall statements apply to multitudinous arrays of processes and structures. The biological primals drive behavior supportive of the necessities of biological life on earth. They serve as an energizing source of motivations and intrinsic human values. They include many genetically prescribed intrinsic regulatory capacities, including overall bodily operational and life support, pervasive chemical integrity, and instinctual overall behavioral directions and regulations supportive of species survival. Many of these operate at the biological level very much as do engineering feedback control systems, in search of target biochemical equilibrium states. Cannon has described and characterized this quality as 'homeostasis'[13].

Processes of adaptive integration sometimes operate as an operative directory cap over the biological primals, utilizing guidance from the valuative representation capacities. It can subsume much of biological primal operations within directives serving adaptation to what we can call a 'situational homeostasis' within external circumstances. It may operate in the small according to intentionality or will; larger volitions and yet larger passive reflective states in themselves are more of the valuative representation capacities.

The valuative representation category--as the home of reflective imagination, representation, perceptual evaluation, and valuation in itself--provides a distinct level of free reflective operational organization. In this it can operate as an independent realm, or in the service of autonomous adaptive direction (personal integration) as an intelligent evaluative guide, or as a superordinate realm, subsuming adaptive integration within free reflective valuation, just as these latter sometimes subsume the genetic preprogramming of the biological primals.

4.6.2 *The neurobiological topography of the inner sensibilities*

The biological primals are rooted primarily in the processes and structures of the autonomic, brain stem, and limbic systems; adaptive integrations are associated mostly with midbrain, thalamus, limbic, and anterior cortical regions; the valuative representation capacities operate primarily in the posterior and deepest regions of all cerebral cortex, and the hippocampus. Many higher functions are globally integrative across anatomical regions and structures, and rely heavily on extensive interconnections across and within multiple cortical and subcortical areas. Consistent with the phylogenetic topography outlined earlier, we see the overall functional organization of the inner sensibilities as mediated across four functional-anatomical regions as follows: rudimentary and autonomic operations (peripheral, spinal, autonomic--dominant in fish and lower); instinctive, appetitive, higher-order sensory (brain stem and midbrain--dominant in reptiles); affective and representational (limbic, midbrain, rudimentary cortical--dominant in lower mammals); highly elaborated valuative representational (cortical and upper limbic--dominant in humans). Organic tripartite functionalism is generally consistent with the levels of construction concept and the broad anatomical topography of the MacLean and Luria views. All include much global interrelations across anatomical structures.

4.6.3 *Neurobiological foundations of the capacities of the inner sensibilities*

The main neurobiological correspondents of the capacities of inner sensibility are roughly indicated in Table 4.1(b). The biological primals are the most genetically predetermined and the most closely related to known neurobiological foundations of the sensibility capacities. These primal operations are mediated largely by neuroelectric and neurochemical processes of the autonomic, brainstem, and limbic systems. They include operational control over many chemical neuroregulators such as endorphins and the hormones, glandins, and other secretions of the endocrine system. Most therapeutic drugs of psychopharmacology act on neurochemical agents of the primals, including limbic catecholamine and brainstem synaptic transmitters which serve appetite, arousal, mood, and pleasure. Distinct local neural networks of these areas mediate highly motivating reward, aversion, and rage. These areas regulate the life-sustaining operations of the autonomic nervous system, and the latter's 'fight or flight' response to threatening situations. Emotions and feelings may likely be associated mostly with limbic neurochemical and neuroelectric activity. Many natural and pharmaceutical chemical agents act on the chemical processes of the brain and body.

Rudimentary instinctual regulatory systems may likely be embodied in fixed action patterns and associated archetypes in the synaptic interconnections and neuroelectric signals of local circuits in the brainstem and midbrain. These may reflect a primitive genetic control of large-scale biologically-significant adaptive behavior which is largely overridden in humans by plastic processes of personal integration more responsive to individual circumstances. Such processes may engage strongly with those of both the biological primals and adaptive integration. Such archetypical templates, and those of perceptual aesthetics indicated below, are genetically determined and may reside through brain stem, mid-brain, and lower limbic and cortical regions.

The operations of personal adaptive integration are likely embodied within neuroelectric signal patterns of mostly anterior cortical, midbrain, and perhaps upper limbic networks and their supportive synaptic interconnections, and their interrelations with other networks and physiological processes. These include significant interrelation with both biological primals and valuative representation capacities. These include a vast array of inner constructions, ranging from simple motor skills to highly complex and abstract context-dependent social circumstances.

In the following discussions all references to the individual capacities (value, aesthetic perception, and apprehension-cognition-truth) refer to the human sense and capacity regarding the capacity mentioned.

The neurobiological foundations of perceptual aesthetics likely include: neuroelectrical signaling in perceptual processes and structures--peripheral sensory structures, midbrain, and posterior cortex; archetypical templates relative to our intrinsic placement in life on earth--see above; and some internal qualities of neurophysiological responses throughout the sensibilities--forms, textures, etc.

Representations of the cognitive-reason capacities are seen as likely embodied in coordinated multi-unit neuroelectric patterns in local and distributed local and composite networks mostly throughout the cerebral cortex (perhaps mostly posterior) and the synaptic interconnection patterns that subserve these. The functional operations of cognitions are embodied in structure and dynamic physiology of vast numbers of synapses which define and sculpt the patterns. 'Synfire chains' and 'sequential configurations' are particular theoretical representations of these kinds of sculpted multi-unit neuroelectric firing patterns. Match-mismatch comparisons of patterns may estimate compatibilities such as coherence, correspondence, or contradiction (Chapter 10).

Manifestations of the human capacity for value are widespread throughout the inner sensibilities as described in the next paragraphs, roughly indicated in Table 4.1(b). Value might associate with personal experience by the mediation of 'significance weighting' by septal input to the hippocampus[14]. The formation of charged or valenced thoughts (chapter three) may take place by similar value weighting of cognitive dynamic patterns in neocortical regions.

4.6.4 *Value and neurobiology*

Human value, like consciousness, acts within all three capacities of the inner sensibilities--the biological primals, adaptive personal integration, and internal representation and valuation. Recall that any of the structural and functional components of the various sensibilities mutually overlap and interpenetrate, and that this is a natural quality of the overall organic organization of the brain and inner sensibilities and their evolutionary origin.

Table 4.1(b). Topographical localization of the inner sensibilities — Part 1.

SENSIBILITY	MAIN NEURAL LOCATION	MAIN NEURAL ENTITY	MAIN VALUE
reflective constructions adaptive integrations	neocortical neural networks (mostly deeper levels)	inter-network connections multi-unit firing patterns	harmonious integrative functioning survival interpersonal & communal rewards & place
biological primals	autonomic system brainstem limbic system	molecular neuro-regulators (hormones, other,...) (cathecholamines, serotonin ..) neuroelectric signals (networks of reward &aversion, ... circuits of the septum...)	bodily vitality, life pleasure, contentment, joy (absence of pain, sorrow) interpersonal rewards
aesthetic perception	sensory neu-al networks, centers of: (brainstem, midbrain, posterior cerebral cortex}	multi-unit firing patterns	aesthetic pleasure texture, proportion, harmony intuition 'just-so' ness
apprehension-cognition	neocortical networks, hippocampus midbrain	neuronal inter-connections; multi-unit firing patterns: sequential configurations; synfire chains	inner representation, construction, reason, validation, truth, arrangement, association cognitive dissonance

Table 4.1(b). Topographical localization of the inner sensibilities —- Part 2.

value	global body physiology neurochemistry neuroelectric activity autonomic, brainstem limbic system, cortex	global neuroelectric neurochemical neurobiological systemic cellular chemistry? molecular physiology?	value in itself meaning, purpose goodness, truth, beauty, justice joy, peak experiences
consciousness	global: body physiology brain metabolism neurochemistry neuroelectric activity brainstem, limbic system hippocampus,cortex	obscure: brain metabolism, neurochemistry neuroelectric currents, fields (fundamental particles or forces)	imagination autonomy awareness

A very wide range of brain and bodily processes, perhaps all, are possible sources of value. Many of one's values can be directly felt in one's consciousness, some strongly, others more vaguely. Also much of value is subconscious. Thus, value can be seen in part as a particular type of conscious experience, and this conscious experience of value, in turn, can be seen as part of value. The dispersive occurrence of our experience of value within neurobiology is conceptually parallel with that of one's consciousness within neurobiology discussed in the first half of this chapter, and parallel with that of one's thinking within something like neuroelectric signals of neural networks, and with that of one's emotions within something like neurochemical distributions in brainstem or limbic regions.

Figure 3.1 has identified three classes of values according to their ground structure as intrinsic, sociopersonal (mixed intrinsic-inherent-plastic), and valenced thoughts. Intrinsic values are rooted directly in particular physiological and neurobiological processes which are genetically prescribed and in their possible further direct associations with other related physiological and neurobiological processes. Sociopersonal values include extensive overlap with other more primitive learned plastic associations many of which we take as inherent to humans (Chapter 10), and with intrinsic values and valenced thoughts. Valenced thoughts are value-charged thoughts which occur mostly as the learned associational weighting (valencing) of individual cognitive elements by value elements. The cognitive elements are likely neuroelectric patterns in cortical neural networks; the value elements are likely processes intrinsically reflective of value in itself or physiological and neurobiological connections with such.

The range and scope of particular values seems exceedingly large. Perhaps very many processes of neurobiology and body physiology have some sort of intrinsic value however subtle, as do the more obvious value-weighted feelings we more readily recognize. Many values seem readily associable with particular neurobiological structures and processes. Elementary biological values of pleasure or rage, for example, seem to be intrinsically associated or one with certain local circuits of the brain stem (lateral hypothalamus) and amygdale, respectively. Others seem more diffuse, global, and difficult to localize. Particularity and weighting of values may vary, but many intrinsic values and the generic dimension of human value are universal. Representative estimates of value are indicated roughly for each main capacity of the inner sensibilities in Table 4.1(b).

The global reach of the value dimension is illustrated by its particular associations with each of the other sensibility capacities. Value shares to some extent this sense of global reach with consciousness. Thus, essential attributes of the individual capacities (for example: pleasure [biological primals]); harmonious functioning [adaptive personal integration], beauty [aesthetic perception]; rationality [apprehension]; goodness [value]) relate to both that capacity in itself and serve the value dimension.

In broad summary, one might speculate with Table 4.1(b) that: harmonious integrative personal functioning is a distinct human value and may be associated with the coordination of multi-neuron firing patterns in and across neural networks primarily of the anterior cerebral cortex and related subcortical regions; that aesthetic experiences of texture or pattern are intrinsic values and may be associated with multi-neuron firing patterns in neural networks of sensory systems and receiving areas of the posterior cerebral cortex; that fundamental biological rewards and aversions are values and may be intimately associated with neurochemistry (serotonin, dopamine, neuroregulators, hormones, and the like) and neuroelectrical activity of local reward and aversion centers of the brain stem and limbic system; that rational processes of understanding and truths are values and may be associated with multi-unit firing patterns and interrelations in networks of the cerebral cortex; and that the peak experiences (perhaps most particularly the mystical religious experience) may relate to widely global neurobiological activities.

Endnotes

1. Keller2000, Ridley1999, Watson2004, Sykes2001

2. Gould2002, Tudge2000, De Wall2001
3. Edelman1987
4. Sarnat and Netsky1974, Noback et al1991,
   Yew et al2001
5. Jackson1932
6. Penfield1975
7. MacLean1970
8. Luria1966
9. MacGregor and Lewis1977, MacGregor1987, 1993
10. Maslow 1968
11. Freud1940, 1977
12. Lacan1966, Taylor1989
13. Cannon, 1944
14. MacGregor1993 chapter 18

Chapter 5

Consciousness and Theoretical Neuroscience

The next three chapters focus explicitly on consciousness itself. This chapter outlines current literature on consciousness and the brain with major representative examples. We see the most significant current work as that which focuses on neurobiological network dynamics, that which pushes to deeper interrelation with underlying physiological processes, and ultimately that which pushes to the interrelations of consciousness with the physical forces and energies by which these higher levels are directed.

Chapter 6 discusses the place of consciousness within the overall world view of science as a whole and considers the possible relation of consciousness to physical theory. Chapter 7 outlines an overall functional theory of consciousness and its relations in brain.

5.1 Introduction and Current Literature

For centuries western civilization has seen consciousness and the soul as the province of the church and theology and embraced a dualist interpretation of their relation with the body[1]. Plato, St. Augustine, and especially Descartes, and then Kant and the British empiricists brought the mind-body relation and the concept of a self as substantive matters into Philosophy. William James established the field of the Philosophy of Mind introspeculatively including widely unlimited descriptions of conscious psychological experiences ranging from daily sensibilities through cognition, pathologies, and paranormal spiritual phenomena[2].

Following reactive logical positivism, this field has progressed in the twentieth century through behaviorism, physicalism (mental states identical with brain states), functionalism (mental states are brain states because of their causal relations), and organic neurobiologicalism.

The most reliable contemporary philosopher of mind is John Searle who stresses the importance of seeing consciousness as a neurobiological phenomenon[3]. Searle has critiqued other current philosophies of consciousness such as those of Chalmers, Dennett, and Crick on this basis. A wayward but highly influential manifestation of functionalism has been the computer information-processing metaphor for brain function. Searle finds this view implausible on philosophical grounds[4]; it has been also derided by neural theorists for decades as we will outline in the last section of this chapter.

Consciousness has become a serious focus in neuroscience only in the last decade. The most important single body of work is that of Gerald Edelman which focuses on the functional and dynamic properties of neuronal groups and the biological grounding of consciousness[5]. This work is outlined in the following section just below. There are also a few recent important experimentally-based contributions[6] from Libet (holistic field theory), Popper (holistic field theory), Tononi (dynamic core hypothesis), Baars *et al.* (consciousness and self), Magistretti *et al.* (consciousness and glial receptors), and others. Integrative Neuroscience is an important coalescence of neural theory which applies the conceptual, mathematical, and modeling techniques of physical science to brain study. This field is currently making important contributions to consciousness study which will be illustrated by the work of Poznanski[7] in the third section of this chapter. Several authors including Penrose, Hameroff, Ramon-Moliner, and myself have recently suggested the relevance of physics and especially modern physics to the mind-brain interface especially because of the free-will question[8]. This area is discussed in the next chapter. The fourth section of this chapter outlines the neurophysiology and physical processes underlying brain operations. Beyond these areas which we see as especially promising, a vast provocative literature has already accumulated in the general area of brain-consciousness study. Access to this general literature can be obtained through our references and their citation lists[9].

There is a natural scale of approach to brain-consciousness study which we could characterize as a series of steps from the bottom-up as: supportive undergirdings, brain physiology, neural architecture, operations-properties, and governing metaphor. The recent literature can be seen as progressively working itself down from the higher to lower levels. The seriously flawed descriptive computer metaphor is being challenged by the recent work on neural networks which is at the levels of neural architecture (neuronal group dynamics) and brain physiology (integrative neuroscience and neural theory). Continued cultivations of these intermediate-level areas, in turn have already made important suggestions regarding the levels of operations and properties (Tononi and Edelman), and supportive undergirdings (physical science-chapter six, and perhaps brain metabolism-chapter seven).

5.2 Neural Networks: Gerald Edelman's view of Neuronal Group Dynamics and Consciousness

Much of the operations of the brain and nervous system is carried out in terms of ongoing sequences of electrical pulses (action or spike potentials) intercommunicated among wide varieties of specialized nerve cells. The functional units are neural networks, often consisting of vast numbers of densely interconnected nerve cells, rather than single cells (neurons). Many neural functions including, say, control of fine-grained muscular movement, or internal representations of thoughts, are thought to be mediated by specific spatio-temporal patterns of these electrical pulses, coordinated across large numbers of cells simultaneously. The nature of such patterns remains obscure, but important preliminary studies have been made in cats and monkeys with multiple microelectrodes[10] and preliminary theoretical concepts of 'synfire chains' and 'sequential configurations' have been formed and computer simulations performed. Recently the term 'coalitions' has been widely used to represent such dynamically coupled groups of neurons. Brain theorists speak of multiple interacting coalitions as the normal mode of integrative systemic brain operations[7]. In the dynamics of large complex networks, as in the cerebral cortex, combinations of excitation and inhibition drive, modulate, and sculpt the patterns and their intercommunications. Recurrent excitation promotes self-sustaining patterns; associated recurrent inhibition sculpts it and prevents it from growing to seizures, which it otherwise might.

There is a long history of interest in this topic[11]. In 1906 Sherrington caught the imagination of the following century of neuroscientists with his image of electrical neural integration in spinal circuits as 'an enchanted loom wielding flashing shuttles--in ever changing, never constant patterns'. In the 1930's Kubie described dynamic 'circularities' in neural tissue. Eccles discovered the all-or-none spike potentials in neurons, and later their precursors--graded post-synaptic potentials--ultimately defining the operative neuroelectrical interaction mechanisms of neurons. Mid-century saw Hebb developing 'cell assemblies' in model neural networks that learned by changing interconnective strengths. Lashley imagined that the billions of cortical neurons could group themselves into a large number of distinct functional systems corresponding to distinct cognitive topics, the same neurons participating in many such systems but in different combinations and permutations. The various functional systems would compete for dominance with selective recurrent excitation and inhibition. He saw specific mental content (consciousness) as associated with the spike patterns and the direction of attention as mediated by the underlying graded general potentials (see the fourth section of this chapter). In the 1960s Moore, Perkel, and Gerstein developed and championed spike train analysis and multiple microelectrode recording from neuronal populations. All these studies provide the backdrop to the development of neural theory (indicated more fully in the next section down) and the current neural network thinking of Edelman and Poznanski.

Edelman's theory of 'neural Darwinism' likens the overall process of cognitive learning in humans by alterations of the effectiveness of vast numbers of interconnective synapses in the brain's neural networks to the adaptive evolutionary selection of genetic alterations to the local environment envisioned by Darwin[5]. His view consolidated findings in neural development which indicated that the brain overbuilds, constructing in infancy several times over as many synapses as are retained. The specific connectivities are constructed under the guidance of an individual's unique genetic inheritance, but the development is winnowed and carefully fine-tuned according to one's unique life experience, primarily during the very critical prenatal and early postnatal learning periods, and continuing especially into adolescence. Large numbers of neurons and especially synaptic connections that are not used die and are weeded away during these periods culminating in a vast harvest in adolescence. Edelman envisioned that only those cells and networks involved in mediating useful systemic operations and interactions are retained. The idea is that in this early harvest one's basic representations and orientations regarding the external world are largely established. Edelman sees one's surviving internally interconnected neuronal groups as having been selected by the winnowing neural development according to their high levels of use and usefulness just as genetic changes retained by a species are seen by Darwin to have been selected by their differential reproductive advantage under environmental selection pressure. Significant learning and modifications continue after adolescence, as do absorption of early learned patterns within larger adaptive plastic systems. Yet, the early years define a significant ground structure with a freedom not so easily available to the later years.

Edelman has written several compelling books on the relation of consciousness to neurobiology[5]. His research has sought the relation of consciousness to especially neural networks themselves, rather than, for example, to certain types of neurons or organelles (see Chapter 7). Lately Tononi and he and others are pursuing this idea in terms of differentiation and unification of network firing patterns particularly in recurrent thalamocortical networks[6].

5.3 Integrative Neuroscience: Multilevel Hierarchy of Brain Physiology

Experimental neuroscience has blossomed particularly in the last quarter of the twentieth-century, emerging slowly well before that mostly within clinical medicine and psychiatry which latter remained largely dominated by psychoanalytic methodology until after mid-century.

Jackson, Penfield, MacLean, and Luria contributed significantly to the topography of brain function (chapter four)[12]. Significant older theory regarding the inner sensibilities has emerged within this psychoanalytic tradition[13]. Freud saw the inner sensibilities as a tripartite complex of: id--primordial biological drives, super-ego--a culturally-learned repressive controller of the id, and the ego--a pragmatic personal mediator between the two. This view sees the inner sensibilities as partitioned against itself. Freud also spoke of energizations in interesting ways, citing tendencies for certain action patterns to accumulate energy which would then prompt their expressive activation. This has similarities to the mid-century views of instinctual behavior in ethology as sequences of 'fixed action patterns'. One of his most significant contributions to brain theory is the recognition of its pervasive subconscious operations. Carl Jung emphasized the prevalence of primitive inherent archetypes in the human psyche which realm he labeled the collective unconscious. Since about 1960 psychiatry has turned increasingly to neurobiological research which produces important understandings, and is progressing in close association with the explosive development of psychopharmacological medications.

Contemporary neuroscience is a highly technical, rigorous, predominantly experimental science, which proceeds in the best scientific traditions of experimental exploration, hypothesis-testing, confirmatory replication, and consensus. It is generally not guided by grand or large-scale theory, but rather works forward piece-meal, across large numbers of laboratories world-wide, on myriad modest ad hoc hypotheses of rather small purview in themselves. The bottom-up approach to the grander questions of brain operations is formidably hindered by the brain's sheer complexity. Moreover, our best experimental knowledge of brain function comes from invasive experimentation which is necessarily restricted to lower animals. These are restricted to the functional capabilities of their subjects and are thus skewed away from the higher capacities of valuative representation and adaptive integration as developed so greatly in humans. Recently a number of multidisciplinary approaches have begun to address these difficulties, many utilizing quantitative modeling techniques[14].

A rich, full approach to neural theory is underway within the newly coalesced field of 'integrative neuroscience'. Integrative neuroscience encompasses the requirements of physical-biological foundations, expansive inclusiveness of scope across biological and psychological variables, multi-leveled hierarchical complexity, and analytical tools demanded by the brain's complexity. Integrative neuroscience embodies the future directions of theoretical neuroscience, and is beginning to provide many bridging recognitions[15].

The main potential of integrative neuroscience is the ability to provide a full predictive foundation for all neural signaling in explicit relation to all salient underlying parameters in terms of the integrated continuous multileveled hierarchy of the brain's physiology. Of special interest to our topic is its further potential to probe and speak intelligently to the long-standing enigmatic relation of brain and mind--it can guide and sharpen the search for the material grounding of consciousness in the brain's electro-chemical processes. These qualities are illustrated in the following outlines of two recent models of Poznanski.

Poznanski's 'Introduction to Integrative Neuroscience' contains both a cohesive overview of recent work in this field and introduces a major new paradigm of the processes and mechanisms of functional organization in the brain[16] illustrated here in Fig. 5.1. The figure applies to any of the hierarchical multiple levels of neural function. The main idea here is that selective self-organizing processes produce physical embodiments of functional organization within and across adjacent brain levels (for example, fields of neurochemical release or multi-neuron coordinated firing patterns) under the control of two types of processes: preexisting (genetically rooted) processes (for example, synaptic transmission), and processes of selection under the influence of ambient environmental variables, labeled 'functional productivity'. The following text taken from Poznanski describes the meaning of this figure further:

Figure 5.1 "...shows the hypothetical hierarchical levels from genesis

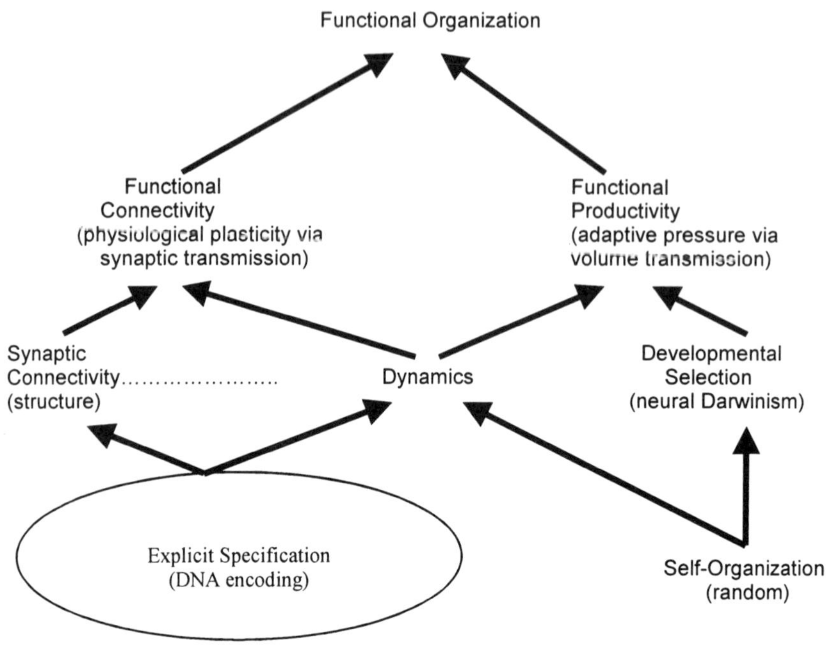

Fig. 5.1 Self-organization of neural functions (from Poznanski[16]).

to function representing a bottom-up hierarchical foundation of Integrative Neuroscience. Structure, dynamics, and selectivity lead to a hierarchically based functional organization composed of biophysical neural networks, indirectly through functional connectivity and productivity during post-ontogenesis and directly during ontogenesis. The emergent complexity occurs post-ontogenetically. The left side of the figure is mostly associated with structural changes (e.g. plasticity, learning), while the right side reflects non-computational processes (e.g consciousness). Note self-organization is assumed to have originated at the molecular level and is therefore observed at higher hierarchical levels to be a random process, but the transition from molecular to cellular hierarchical levels invokes a non-computational or nonalgorithmic process that is not necessarily random. The dotted lines between synaptic connectivity, dynamics, and developmental selection indicate ontogenesis. The cycle is repeated with explicit specification in the offspring via DNA encoding, as indicated by the dotted circle. Explicit specification remains the only brain operation that is implemented computationally."

Poznanski's theory of cognition in composite cerebral neural networks[17] is particularly representative to our purposes. This theory sees cognition represented in semantic meanings associated with dynamic activity patterns of multiple interconnected coalitions of neurons (neuronal networks), and particularly involving extrasynaptic influences mediated through the extracellular fluids of the brain, and explicitly engaging multiple levels of the neurobiological hierarchy indicated in Table 4.1(a). This allows for the influence of cognitive processing by ambient brain environment, and through this, Poznanski suggests, by consciousness. Poznanski sees consciousness as reflective of integrations across this material hierarchy, especially associated with volume transmissions, and with semantics.

Poznanski's thinking is a good contemporary example to help push us closer to the foundational obscurities of the neural-experiential interface and the enigma of apparent partial conscious autonomy as described in chapter six. Particularly significant is his identification of volume transmission as a pervasive medium of influence on neuronal behavior outside of conventional anatomical synaptic interconnections--the path he calls 'functional productivity', after Globus, and associates with consciousness. These pathways are dependent on body physiology and allow a wide mix of ambient variables to influence neural activity.

Equally significant is the association of consciousness with glutamate receptors of glial cells (following Magistretti and Pellerin[6]) and therefore possibly with the brain's metabolic system (see Chapter 7). His localization of consciousness is somewhat similar to Lashley's, but goes deeper into its neurobiological hierarchy. Note further that within such an explicit physiological framework, any extra-neural action of consciousness such as partial autonomy can be more clearly considered in terms of the brain's physiology, and the physical forces and energies underlying this physiology. As we will take up in the next two chapters, this approach suggests that developments in theoretical physics, exotic or not, may provide a possible middle ground for resolution of the ancient mind-body quandary in twenty-first century terms.

Integrative neuroscience is a restructuring and expansion building from a long history of analytical theory and modeling of the functional organization of the brain and nervous system by physical scientists, engineers, and mathematicians. 'Neural theory', where the convenient label is taken as inclusive of all such efforts including integrative neuroscience, has succeeded in establishing and continues to improve mathematical frameworks for describing the physical laws governing neurobiological operations and signaling, and has explored the operations of many particular neural subsystems throughout the animal kingdom.

As in physical science generally these approaches move beyond the levels of description, associations and correlations, and the lateral analogical thinking of metaphors to that of more precise characterizations of underlying causal agencies and parameters defining neural systems. Neural theory has four main qualities: (a) it is rooted in the complete hierarchical complex of the brain's physiology (indicated in Table 4.1(a), (b) it is, through these, relatable to the fundamental governing principles, mechanisms, forces, and energy sources of the brain's operations, (c) it is, through these, characterizable in the mathematical terms, methodologies, precisions and discerned approximations of the physical sciences, and (d) it is, through all these, able to provide undergirding for the guidance of speculative thought and experimentation, and the interpretation and prediction of experimental findings. An outline of the history of neural theory is given at the end of the next section.

5.4 Neurophysiology and Physical Processes

The foundations of neurobiological functioning and its roots in physical processes are contained within its systems of electrical currents of flowing charged ions and its systems of chemical transmissions and regulations, as embedded in the biophysically supportive and nurturing extracellular media of the brain[18,19]. We will briefly outline these in terms of neuroelectrical signals, ionic fluxes, and neurochemical processes.

5.4.1 *Neuroelectric signals*

The output neuroelectric pulses of neurons mentioned above are generated in neurons in response to smaller graded electrical fluctuations incurred in the neurons in response to input signals. Each excitatory and inhibitory input produces small positive or negative electrical fluctuations, respectively. The accumulative fluctuations of multiple inputs and across small time periods (milliseconds) are called generator potentials. At certain points in most neurons, large, brief, all-or-none spike potentials are triggered by the generator potentials. (Usually these trigger according to a malleable threshold rule: if the generator potential exceeds the threshold, an action potential is triggered.) Spike potentials typically come in groups according to the accumulation and temporal spreading of generator potentials.

The electrical currents which underlie the electrical potentials travel in closed loops which traverse neural membranes and flow through both the internal and surrounding external fluid regions of neurons. Electrical potentials can therefore be recorded either within or external to cells. Further, such recordings can be made with either very small microelectrodes, recording signals from one or several cells, or larger macroelectrodes which record collective signals from large numbers of local or regional cells. Macroelectrode recordings (extracellular) are useful for rough estimates of global behavior. Extracellular recordings are useful for spike trains of individual or a few cells. Intracellular recordings are useful for both individual spike trains and intracellular generator potentials.

5.4.2 *Ionic fluxes*

The electrochemical functioning of the nervous system and the inner sensibilities depends upon the well-being of the biological support system in which it operates.

The electrical signals of neurons are fluctuations in the electrical potential across the narrow membrane which forms a neuron's external wall. This potential is maintained in all living biological cells by the maintenance of a selective difference between the concentrations of ions in the internal fluid of the cell and those in the surrounding external fluid. The surrounding fluid, like sea water, is higher in salt: sodium (Na^+) and chlorine (Cl^-); the internal fluid is higher in potassium (K^+). These ionic concentration gradients are maintained in living cells by metabolically driven molecular pumping agents in cell membranes. These processes are in the realm of molecular biology in which many secrets of life remain yet obscure. (For example, the mechanics of the organization of repeatable order within proteins and macromolecules like DNA, and its molecular construction of living bodies and all their physiological processes is a foundational question in molecular biology, comparable to the mind-body question of neurobiology and the inner sensibilities. One can imagine that the known physical forces operating among the smaller constituent molecules--which should be primarily electromagnetic--could bring this about utilizing reliable elementary patterns of geometrical structure, but its likelihood challenges our credibility. This question is considered again in Chapter 6.)

Neurobiological functioning depends upon the metabolic maintenance of the molecular ionic pumping agents in nerve cell membranes, and general well-being of physiological state. These, aside from the deeper molecular obscurities, are understood in terms of known physical processes operating within the realms of known physical law.

5.4.3 *Mechanisms of neuroelectric signaling*

We can now consider the particularly neurobiological processes underlying neuroelectric signaling. The initiation of a fluctuation in the generator potential of a receiving cell is typically initiated by the release of a chemical transmitter at the input terminal of the sending cell, which in turn traverses a small gap of extracellular fluid to briefly open selective pores in the membrane of the receiving cell. This allows particular ions to flow into or out of the receiving cell according to their concentration gradients and the local electrical gradient. These flows of charged ions are electrical currents. The ions flow in continuous closed loops through the internal and external regions of the cells inducing and including transmembrane flow of ions and the graded rotations of the electrically charged lipoprotein molecules which comprise the neuron's membrane--all in lawfully coordinated fields within the altered concentrations and electrical gradients they produce. These ionic currents and the corresponding electrical gradients and transmembrane voltage drops are the neuroelectric signals that the electrodes measure.

The motions of the ions are subject to the well-known determinative physical laws of electrochemistry, which are traceable ultimately to the fundamental laws of motion of charged particles acted upon by all known physical forces including particularly electricity, interparticle collision, and inertia, constrained here according to the structural features of the neuronal environment. These ionic current flows are channeled and guided by the anatomical particulars of the surrounding neural structures and the intra and extracellular fluids. There is a rich variety of neurobiological mechanisms which influence neuroelectric signaling. All of these manipulate these diffusions of ions and are ultimately dependent upon the physical forces which drive their motion within the surrounding constraints as indicated above. The extracellular fluid contains ions, other molecular particles, and cells all of whose physiology is also controlled by similar electrochemical diffusion laws and molecular gating. This body physiology underlies the overall state of neural functioning and otherwise influences it in obscure ways.

5.4.4 *The neurochemical dimension*

Functional neurochemistry[20] entails five or six families of neurochemical transmitters which mediate synaptic transmission of electrical signals as outlined above, and a large variety of more obscure neuroregulators. Chemical transmitters seem to reveal an interrelated duality in neurobiological control. On the one hand neurotransmitters may be seen as simply varieties of chemicals that serve the common realm of neuroelectric signaling, different chemicals being used at the outgoing synapses of different neurons (for example, acetycholine or glutamate transmissions). On the other hand, different chemical transmitters are associated with different particular functions. For example, the family of monamines in the limbic system--epinephrine, norepinephrine, dopamine, serotonin--are selectively associated with states like well-being, arousal, pleasure, mood, and sleep cycles, as well as serving particular synaptic transmissions. This association implies a level of selective chemical functionalism in the inner sensibilities. Given drugs can be used to influence certain functions and certain syndromes such as anxiety, depression, Parkinson's disease and so on. Probably all can be reduced ultimately to physical forces acting on neural ionic mechanisms described above and mediated by intra and intermolecular forces of chemistry (mostly electrical and electromagnetic).

Neuroregulators are released at output terminals of nerve cells by their neuroelectrical signals as are transmitters, but diffuse more widely and globally, acting over long distances and long times. Their effects are correspondingly more diffuse and global, and often more biological than neural--pain suppression, control of hormone release, influences on the life-supportive processes of the autonomic system. All these can be seen as subject to the known determinative intra and intermolecular forces of chemistry.

5.4.5 *Neural theory and neurophysiological processes*

A central feature of neural theory is the construction of a mathematical formalism to relate spatiotemporal relations among the intracellular, extracellular, and transmembrane electrical currents and potentials in neurons in terms of structural and physiological parameters including especially the selective fluctuations in transmembrane conductance (permeability to certain ions) which represent the central natural driving mechanisms of neuronal signaling[18]. Much of this formalism is based on a simple electrical circuit model which represents very clear physiological attributes[21]. The model and its formalism can be applied to local regions in neurons (eg, the cell body), to stretches of dendritic membrane (Lorente de No, Rall), to the generation of action potentials (Hodgkin and Huxley), or other endogenous active membrane signaling, the activation of local neuronal regions, composite neurons as collections or integrals of unitary regions or approximated as 'point neurons' representing the generations of spike trains in the soma. Networks of neurons can be represented by interconnecting such representations of indvidual neurons. Hundreds of such studies have been performed all over the world on neural networks of all kinds over the last half century[22]. This theory focuses readily on transmembrane potentials which can be directly compared to intracellular records. It can also be used to describe extracellular currents and potentials. Particular applications can become complex but can be formulated and computed. Principles and tendencies can sometimes be drawn more readily. Neural theory has also included the development of spike train analysis for single and multiple spike trains in neurons and explored the nature of correlated activity in variously connected networks of both computer-modeled and real neurons[23]. Neural theory has applied dynamic similarity theory to neural networks. More recently, integrative neuroscience has increased focus on neural hierarchy and volume currents (Poznanski), multilevel system equations for neural activity (Chauvet), non-computational cognition (Globus, Kercel *et al.*), metamodeling (Lewis), extracellular currents (Costalat and Delord), neuronal cable theory (Lindsay, Rosenberg, Goldfinger) and other important topics[24]. Several overall reviews of neural theory are available[25].

5.5 The Computer Metaphor for Brain Function

The computer metaphor for brain function refers to that body of scholarly and research activity rooted in computer science and artificial intelligence which interprets and models the operations of the nervous system and brain as if it were a computational information-processing system. Neural and psychological signals are taken as ultimately discrete; operations are essentially cognitive and informational; all operations are 'computational'; organization is quintessentially functional, algorithmic, and modular; sensory operations effect pattern recognition; and so on. Extreme views are soberly considered and widely held among many adherents, including, for example, the idea that consciousness is a property of some larger class of information processing systems so that certain kinds of robots might well be conscious.

The computer metaphor has dominated the field of brain and neural network modeling in both the popular press and many neuroscience research groups for the last twenty years or so. The willingness of the public and especially so many experimental neuroscientists to be taken into this metaphorical view of the brain surprises and dismays neural theorists and many others who have taken a fuller consideration of the brain's natural physiological processes. Although a simplified computer-like model can provide a limited useful interpretative view of part of the dynamic signaling of some brain networks in certain circumstances, the overreaching application of such partial models as a realistic view of the behavior and nature of the brain as a whole is completely unjustified. This rampant overreaching is precisely the error of the field and its credulous endorsers.

The modern substantive literature of the computer metaphor for brain function[26] originated in the 1943 paper of McCulloch and Pitts which grounded the binary information-processing paradigm in an analogy with the then recently discovered all-or-none spike trains of neurons. Adaptation and feedback control systems described by Norbert Wiener and elaborated in automata theory founded the highly influential dimension of adaptive cybernetic control to the artificial intelligence view of behavior in the post-WWII years. Hebb's theory of learning in neural networks (1949) as cell assemblies with selectively enhanced synaptic strengths and the elaboration of such properties by many modelers such as Rosenblatt (1962) initiated the field of adaptive computational neural networks. In the 1950s and 1960s artificial intelligence was defining itself with a strong sense of importance for the cold war, and the associated philosophy of computer intelligence, including the famous Turing test for 'thought' in computing machines became prominent. In the 1970s, the field was adopted within cognitive psychology groups all over the country forming the restructured and heavily AI- and computer-oriented field of 'cognitive science'. In the 1980s a second and highly pervasive wave of 'computational neural networks' surged providing many kinds of 'smart' and sophisticated adaptive 'hi-tech' electronic control modules with many widespread practical applications. In the 1990s the computational view of brain function, with its discrete variables, algorithms, and modules became sacrosanct. The present decade has seen the field turn to consciousness study. Two current representative and complementary statements of the computer metaphor are the overall discussion of the current computational litany, *How the Mind Works*, by Steven Pinker, and *The Quest for Consciousness*, by Cristof Koch. These books, though entertaining and useful for limited purposes, are, like the field they represent, fundamentally flawed and misleading as overall models of the human brain and mind.

Neural theorists have long voiced critical views of this field. The computer metaphor is groundless--resting on an overly reduced interpretation of brain operations, an unjustifiable analogy, and hollow labels. It allows only lateral, analogical thinking. It incorporates only the surface level of the intrinsic multi-leveled hierarchy of the brain's neurophysiology and its natural continuous physiological processes. It has no relation to underlying causes nor any predictive abilities. It gives no guidance regarding the ultimate nature and relations of consciousness in brain. The view imposes a biased, self-limiting overreduction of human consciousness restricted to cognitive information-processing terms, whereas in fact, and this is perhaps the blindness of the computer metaphor, our extracognitive sensibilities are in themselves much more than our secondary conscious cognitive representations and interpretations of them. These sensibilities manifest in consciousness from awareness in itself and reverie, through any variety of basal or peak experiences. At best, the computer view gives us a superficial picture of neural operations and a diminutive view of our own nature, falling seriously short of the essential natures and ground stuff of both the brain and consciousness. At worst, it leads us into misguided views of brain operation such as 'effortless manipulation' which seems to fly in the face of the second law of thermodynamics with no mention of the deeper problems involved, and offers us the silly idea that robots are conscious. All around, the mindset of the computer metaphor is better geared to artificial intelligence than to the human brain. It is damagingly misleading when applied to the human brain.

The most hopeful view of all this is that it seems that the computer metaphor is beginning to pass out of favor by the progression of understanding of this natural subject matter as indicated in the first section of this chapter. The shortcomings of the field are becoming more obvious to increasing numbers of researchers and students of both consciousness study and neuronal network dynamics. Gerald Edelman has provided a broader integrative perspective[27]. Roman Poznanski has provided a pointed and comprehensive critique of computationalism as a model of brain dynamics[16]. John Searle has argued against the view on philosophical grounds[28], labeling it as 'utterly' and 'wildly' implausible[29]. I have pointed out the weaknesses of the field over four decades[30]. Other current brain theorists have also recognized these and other limitations[31].

Finally, at a broader historical level, consider the Turing test which asserts that a computing machine can 'think' if a human observer cannot distinguish its replies to questions from those of another human[32]. It is notable that this test, published in 1950, is immersed in the same 'black-box' methodology of behaviorism and some systems theory which gave birth to the 'Skinner boxes' for programmed conditioning of human infants. This is a groundless kind of thinking which focuses on lateral relations between input and output variables of a system' (ie, a black box) in themselves without regard for its inner operations or causal agents. Both Turing and Skinner's approaches and their parent behaviorism are all reflective of the hyper-analytical logical positivism which flowed from London and the Vienna circle of the nineteenth-twentieth *fin de siecle* through the following half-century[33]. The Skinner boxes have long since given over to more fully human concerns, exemplified by current interest in nurturing and empathy (Chapter 14). The computer metaphor for brain function still overreaches far beyond its fitting limited place within the fuller natural scope of its subject matter.

Endnotes

1. Edwards1967
2. James1890, Flanagan and Kim1995
3. Searle1984, 1992, 1995, 2000
4. Searle1997
5. Edelman1987, 1989, 1992, 2004
6. Libet1994, Popper et al1993, Tononi and Edelman1998, Baars et al2003, Magistretti and Pellerin1999
7. Poznanski1999, 2001, 2002
8. Penrose1989, 1994; Ramon-Moliner1994; Hameroff1994; MacGregor2002, 2004a, 2004b
9. Crick1994, Dennett1991, Koch2004, Porter1994, Rosenfield
10 Gerstein et al1978, Abeles1982
11.Sherrington1906, Kubie1930, Hebb1949, Lashley1960, Moore et al1966, Gerstein and Perkel1969
12. Jackson1932, Penfield1975, MacLean1970, Luria1966
13. Freud1923, 1940, 1949, 1977; Ethology; Jung1953-61
14. Cappa2001, Delacour1994, Reggia et al1996, Hasty et al2002, See also the Journal of Biological Psychiatry
15.Globus1992, Searle1992, Chauvet1996, Gordon2000, Poznanski2001; see also the Journal of Integrative Neuroscience
16. Poznanski2001
17. Poznanski2002
18. MacGregor and Lewis1977, MacGregor1987, 1993
19. Eccles1957, 1964
20. Bradford1986, Yusuf1992
21. Eccles1957, 1964; Hodgkin and Huxley1944; Davis and Lorente de No1947; Rall1977
22. MacGregor1987, Poznanski1999, Reeke et al2005
23. Moore et al1966, Gerstein and Perkel1969, Gerstein et al 1978, Abeles1982, MacGregor1972, 1991
24. Chauvet1966, 2002; Globus1992; Kercel et al2003; Lindsay et al2005; Lewis2003; Poznanski2002; Costalat & Delord2004; Goldfinger2005 Journal of Integrative Neuroscience, www.worldscinet.com
25. Harmon and Lewis1966, Freeman1972, 2000, Jack1975, MacGregor and Lewis1977; MacGregor1987, 1993; Poznanski1999; Reeke et al2005
26. McCulloch and Pitts1943, Wiener1961, Rosenblatt1962, Gardner1985, Gluck and Rumelhardt1990, Pinker1995, Koch2004

27. Edelman1989, 1992, 2004
28 Searle1992, 1995
29. Searle1997
30. MacGregor and Lewis1977, pgs 322, 384-388;
 MacGregor, 1987, 1993, 2004
31. Penrose1989, 1994; Globus1992
32. Turing1950
33. Edwards1967, Flangan and Kim1995, Kramer2002

PART B

AN INTEGRATIVE PHILOSOPHY OF BRAIN AND CONSCIOUS EXPERIENCE

Chapter 6

Consciousness, Physics, and Neurobiology

The relationship of brain and mind is one of perhaps the three greatest questions of science, the other two being the question of the ultimate origin and energization of the universe, and the nature of the energization establishing and supporting the highly ordered coding of the DNA molecules and the corresponding multitude of structuro-functional order it brings into being through what we call life. Western philosophy has struggled with the 'mind-body' problem in essentially modern terms for a century or more, but has not brought us near its obscurities[1]. The preceding chapters have argued for the centrality of conscious experience and its qualities to our organic functioning as well as to our existence. As individuals, our primal sense of being is in conscious experience, in the passive detached reflection we experience with it, and the responsive partial control it gives us over our neurobiology. We know our neurobiology is central only when it is out of balance.

6.1 Texture, Interfaces, Energization, and Nature of Consciousness

The central ingredients of the mind-body question are those of texture, the consciousness-physical interface, and, energization and causality.

6.1.1 *Texture*

A moment's reflection convinces us that the elements of our conscious experience are of a different nature entirely from the physiological processes of neurobiology, that we have no direct conscious impression of neurobiological process as such, and that if we are to think of them at all we must imagine them as we do other hidden qualities of external reality, secondarily, externally, and largely through collective communal study.

The content of our conscious experience and operations is of the nature of awareness, judgments, attitudes, images, moods, emotions, values, intentions, inclinations, apprehensions, cognitions, feelings and the like. In neurobiology we may speak, for example, of coordinated neuroelectric patterns, neurochemical transmitters, neuroregulators, neurons, glial cells, receptor molecules and so on. In considering operations we can contrast subjective states of well-being with neuroelectric activation of the lateral hypothalamus or distribution of serotonin; or, say, coherence or contradiction (cognitions) with, for example, presumed corresponding match-mismatch of dynamic neuroelectric patterns in neural networks. All these entail a clear distinct gap in texture. The association of the contents of consciousness to their apparent neurobiological elements is a bald assumption across a physical-experiential interface which we cannot clarify until we better understand the nature of consciousness and its fuller relations within the brain. In this assumption we have representations on both sides of a textural-modal divide standing in presumed association with each other.

Further, as described further in chapters eight-ten, the direct conscious experience of a cognitive association or idea in itself may be quite different from the conscious experience of its expression in language. The need to let some realizations 'sink in' may relate to this, as is clarified in the next section below and following chapters.

6.1.2 *Interfaces*

It is important to recognize that any particular assumed association between consciousness and neurobiology reduces to an apparent correspondence. Consider, for example, one's simple passive cognitive experiential sense of a given visual image. Presumably this image corresponds to neuroelectric signals in various regions of one's brain. No matter how precisely we might identify that neuroelectric (or other physical) basis, the experiential image is always a clear mental reality quite distinct from the neurobiology. Even an explicit identification of a neurobiological determinative correlate does not take us across the textural gap between the physical and experiential realities.

This obscurity can be seen to remain when one considers slightly higher-level entities such as another living creature, which presumably will be represented in one's brain by a wider range of neurobiological particulars such as multiple sensations (visual, auditory, olfactory, tactile), movement, anticipations, categorizations, and so on. There is a definite consciously perceived entity and a corresponding realm of things called imagination within which that entity plays a part. This imagination may well be thought to relate in some global sense to a very complex mass of neurobiological (probably neuronal) interconnections amongst many particular and higher level entities and relationships. Here also, even if one could identify the entire array of physical connections and signals involved, the textural gap between the imaginative realm and the physical is still not crossed. The imaginative realm is a distinct mentoexperiential reality (centrally conscious or potentially conscious but also organically involved with some uncertain degree of subconscious content) quite different in manifestation from the presumed physical signals. It would seem the only satisfying way to bridge this apparent disjunction would be to go beyond mere association of mentoexperience to neurobiology to the successful identification of consciousness as one with some physical process. This last step, however, should be explicitly recognized as an hypothetical substantive monism of identity beyond our limited functional monism. In this case the conscious experience could be taken as a quality of that process, but the disjunction and obscurity would yet remain.

It might seem silly to have such experiential images and imaginings, be they simple explicit visual images or more complex or abstracted relations of higher-order entities, if neurobiology in itself is sufficient undergirding for all our sensibilities. To see that it is not a silly situation requires only for us to picture our active autonomous actions.

6.1.3 *Energization*

We experience the ability to consciously direct both our attention (which takes place within fields of imagery and imagination), and the movement of our body in external reality. One must suppose that either class of autonomous act must somehow induce neurobiological alterations at some locus or loci which in turn bring on the fuller desired changes in mental imagery (attention) or physical position of body.

The only other alternative to this supposition is that our autonomy itself is an illusion, that our whims and deliberate decisions, as well as drives and impulses, are in fact determined by the entire field of neurobiological (and physiological signals) themselves, all behaving in accordance with physical law. This, I take as too far out of line with our direct sense of existence in our selves and local environment, and of the behavior of other people and creatures to take seriously. Further, to have such mental imagery and senses of autonomous influence all as passive self-deluded riders to determinative physicality seems indeed too silly. Yet further, it is not easy to see any value for such an ineffectual conscious experience and false autonomy. Finally, to say that the neural does it all, that partial autonomy is an illusion, is quite like saying (before string theory) that gravity didn't exist because it hadn't a place in the standard model of physics (see below).

From all this our partial autonomy appears as the most revealing manifestation of consciousness. Our directly felt ability to direct our attention or move our body, which translates into directed alterations of neurochemical and neuroelectric states or signaling, requires a force. Energization is the ability to exert a force, and is here the manifestation of causality. In this case, what we are (consciousness) may be best indicated by what we can do (partial autonomy). Moreover, to see autonomy, which is free will, in terms of energization brings us closer to a realm at the heart of physicality and physical law, and may thereby reveal some foundational common ground for bridging the gap between the conscious experience and neurobiology.

6.1.4 *The fundamental nature of consciousness*

All of this may help us understand a possible reason for the obscurity of the textural gap between the consciousness and the physical. To avoid the apparent reductio ad absurdum--that is, to allow even partial autonomy within the realm of determinative physical law--is it possible, even necessary, that consciousness maintain a distinct region of identity beyond what can be identified within physics as we know it? This would seem to be the case if we want to maintain both partial autonomy and the determinism of physical law that seems to so pervasively govern our macroscopic physical universe.

Separately, is it not possible that consciousness in higher life might be especially useful precisely because of a capacity to free local self-direction from clumsy, cumbersome determinative physical law in idiosyncratic local circumstances? Yet further, if this is so, does this require consciousness to be partially non-physical in the sense we currently know physicality? Is it possible that microscopic uncertainties of quantum mechanics in some of the brain's particles become operationally linked to neurobiological effects (transmuted, as well be said here) thereby opening and passing through a door of indeterminism to help guide behavior and further then progress to superordinacy? If so, how would this relate to the non-physical dimensions of consciousness? Or, would this mean, the term non-physical would no longer apply? One would need more specifics regarding physical linkages and mechanisms of action.

Such alternatives could introduce and describe the characteristics of certain types of indeterminism localizable by some means to regions of the brain (or biology) or the identification of a new type of causal force beyond those that have been so far recognized in physical theory, again with propensities to act exclusively or especially in regions of the brain (or biology). If this is so, how may we best modulate or otherwise explain our vision of the physical in relation to consciousness? The existence of some indeterminism in itself would not be enough, unless it contained additional explicit guidance as to the operations and nature of consciousness. These questions are considered further in the following sections and the next chapter. (Recent books on consciousness and the brain are listed in the reference list).

6.2 Conscious Experience and Physical Science

6.2.1 *The grand unifying view of physical science*

Modern science sees the whole of things as consisting of three layered strata--the physical, the biological, and conscious experience. Each of the three strata of things--physical, biological, and mentoexperiential--has its own self-relevant patterns of activity, organization, and dynamics. Each stratum has its own mysteries of origin and nature. Especially, each stratum has a foundational mystery of energization. Herein--in these obscurities of energization deep in the organizational hearts of biological life and the mentoexperiential --reside the kernel questions of the science and fundamental thinking of this new century.

Most scientists believe in the primacy of the physical in the sense that: the physical is continuous and organically interrelated throughout vast regions of space whereas all known life or conscious experience has been found only on earth; all life we know of is associated with its own particular structural forms of our common earthly sampling of apparently universally common physical matter; all conscious experience we know of is associated with ourselves and likely some other uncertain range of lower life forms. Most contemporary scientists believe that the mysteries of life and of consciousness have evolved or emerged within physicality and eventually will be understood in relationship to that physicality.

It should be clear that if this is the case, it will in no way diminish the richness, complexity, grandeur, nor preciousness of experience, life, nor the whole. Physical science is already a very rich, complex, and grand creation in itself. If it evolves far enough to include life and experience it must necessarily come to recognize the full richness and complexity of these--no incomplete model can be adequate.

Moreover, it is still far from certain how this view in science will fare. There are many deep complexities, and several foundational obscurities which may remain for us ultimate mysteries. There is danger in preemptive overreduction and disregard of the full dimensions of these.

6.2.2 *Molecular biology and life, and consciousness and autonomy*

Two very successful fundamental sciences have blossomed in the last half-century: that of the deep structure of matter (the 'standard model' of quantum physics and its recent formulation in string theory) and that of the primordial and universal deep organizational foundations of life in the macromolecular molecules which make up DNA and proteins. In the grand physical view of the universe, these two sciences are the foundations of the first two strata and speak directly to their obscurities of origin and nature. Consciousness may likely play that same role for the third stratum in its relationships in the brain's neurobiology. Life and consciousness might be characterized as 'emergent phenomena' since the origins of the upper two strata may be so characterized.

The highly significant developments in twentieth-century molecular biology suggest the picture that all life on earth (except perhaps the most extremely primordial) is founded on a primordial and universal entrapment of particular structural arrangements of the elements of certain compound molecules, which in themselves produce or direct at least these following three structuro-physiological processes: they produce and utilize energy from surrounding matter (by chemical reconstructions); construct their own anatomical structures and produce their own physiological processes; and replicate themselves.

There is considerable order in the physical universe[2], but life forms are primary originators of order on earth. Nonetheless, the second law of thermodynamics (physical science) predicts the continual degradation and ultimate loss of order into uniformity of the universe as a whole and of any portion of it sufficiently isolated to qualify as a 'closed system'. The production of order by life forms will not violate this law because living organisms are open systems needful of ambient matter and energy. Physical science says the larger portions of space containing the local order-production degrade more order than the enclosed life forms produce, when energy and matter influxes are taken into account. Incident radiant energy from the sun is sufficient violation of the isolation requirement to negate the 'closed system' assumption.

The word 'entrapment' used above reflects the molecules' abilities to maintain, effect, and reliably replicate these three structuro-physiological processes. This entrapment must be mediated by forces.

It is generally assumed that such entrapment is brought about by traditional forces acting according to known determinative physical laws, probably in conjunction with particular structural arrangements. Since the structuro-processes involve molecular recompositions, traditional electromagnetic forces are very likely involved. Further, since electromagnetic forces both originate in and, in turn, act upon, the electric charges and their motions in these molecules and elsewhere, it is plausible that these foundational processes of life, and therefore life itself, could be mediated and sustained by the physical laws of interactions between electrical charges (fixed and moving) and electromagnetic forces, entrapped in these highly orderly and active complex arrangements.

This entrapment, then, in principle could resolve the second of two central mysteries of 'emergence' with the current views of creation in physical science--that of the emergence of life within physical processes and physical law. However, this supposition and this area are very uncertain. It would seem that satisfactory support of this view hinges on convincing demonstration of the origin of such molecules within the natural operation of known physical forces within naturally-occurring circumstances in the evolution of the universe, either on or near the earth or distant in space (especially the DNA molecule which serves as a mother template for all the material construction and all the physiological processes of every living creature on the earth--except perhaps the most primordial if these latter are to be considered life in the same sense[3]). Moreover, biologists still face the very complex mechanisms through which large molecules like proteins carry out the vast and complex structural constructions and physiological functions by the interaction of natural forces and geometry. Both tasks are daunting.

The second obscurity of origin (or emergence) is that of consciousness, with its autonomy and other enhancements of being. We don't know where to localize these. Are they associated particularly with the plastic autonomous experience and directions of humans and their recent antecedents? Or do they emerge somewhere earlier: It is plausible to suggest the emergence of consciousness or some other obscure quality near the fundamental levels of life rather than in the complexity of the brain? To what degree, if any, do consciousness or partial autonomy associate with overall guidance of behavior in gorillas, cougars, crocodiles, spiders, or flatworms, jelly fish, coral colonies, or algae? Are they associated with the primal organizations and directions of primordial molecular life like viruses? Or do consciousness and partial autonomy originate somewhere deeper in the physico-energetic stuff of the universe itself?

Although neither autonomy nor consciousness is apparent in inert (non-living) matter, some degrees of order and energization certainly are, in both microscopic and macroscopic realms. Yet further, the molecules of life and of neurobiology are constituted of the primal physical stuff of the universe. So the search for roots of autonomy and consciousness in foundational physical theory is also plausible. Wave-particle duality and quantum mechanical uncertainty may play some part, but are seen by many in theoretical physics as limited to the microscopic realm, largely if not entirely, buffered from our mesoscopic life and neurobiology. Yet this claim may be less secure at certain places where atomic microstructure is particularly close to physiological processes as we will show below.

One may speculate that consciousness and autonomy have emerged according to Darwinian natural selection because of the usefulness of their apparent two-way mutual interaction with or within neurobiology in the evolution of open-ended control of individual behavior, especially higher neural plasticity. This picture might suggest an association of consciousness and autonomy (or at least a great flowering of these) with more recent physiological levels rather than much older universal macromolecules. This would also suggest a natural buffering and constrainment of consciousness and autonomy within larger genetic controls and ambient surroundings, which we do indeed see. For example, the stronger relations of consciousness and autonomy to adaptive personal direction than to autonomic processes of life-support would be consistent with this.

6.2.3 *Theoretical physics: Quantum mechanics and string theory*

Current theoretical physics revolves around the so-called 'standard model' of quantum mechanics which explains a vast range of physical behavior. The standard model is mathematically rooted in the

Schroedinger wave equation (expanded now to represent string theory) which predicts the occurrence and interchange of energy and matter in the form of waves and particles (which latter may be either matter or forces, depending upon their 'spin'). The model is impressively successful in predicting the existence of all subatomic particles and through these all observable matter, and the dynamic behavior of all matter from these smallest elementary particles to the vastest regions of cosmic matter from four known forces: the strong , the weak, electromagnetic, and gravity. Quantum mechanics in itself has at least three foundational dualities: wave-particle, force-matter, and matter-antimatter. The uncertainty principle might seem a promising foundation for autonomy. It tells us that even in quiescent empty regions of space, the universe, on the microscopic scale, is both intrinsically uncertain and frenetically active in interchanging local quantities and forms of energy and momentum. According to quantum mechanics, however, none of these dualities, nor the uncertainty principle would affect traditional classical physical laws above operative distances of a nanometer, 10^{-9} meter, or less, more than a thousand times smaller than the nominal micron-level of neurobiology.

Quantum mechanics has successfully generalized itself to include Einstein's special theory of relativity which deals with electrogmagnetic forces, light, time dilation, and speed compression. With the help of string theory it has absorbed his general theory of relativity, which represents the force of gravity by curvature of space and therein seems to successfully describe the mechanics of the universe through all time back to at least close to its origin some thirteen billion years ago. String theory also holds the promise of resolving several fundamental problems of quantum mechanics, such as its failure at points (because the mathematics break down from divisions by zero distances), providing smooth elegant predictions of several assumptions in quantum mechanics, and predicting numerical values for parameters that were previously determined only by experiment and ad hoc postulates. String theory holds that the essential physical ingredient of the universe is a closed loop of string (whose diameter relates to a total energy value and has been estimated). Every matter particle and every force carrier particle is a certain pattern of vibration of this string. Space is higher dimensional (current thinking is that there are eleven dimensions). The mathematics is exceedingly difficult. Researchers estimate the development of the theory to occupy some half of the coming century. Excellent qualitative introductions are given to quantum mechanics by Pollack[4] and string theory by Greene[5].

6.3 Two Speculative Views of Quantum Mechanical Consciousness

The last section of this chapter presents a map of possible views of the relationship of consciousness to the brain by contrasting its existential and physically-based manifestations. First, though, this section outlines two instructive quantum mechanically-based views of consciousness and the next two sections give a more comprehensive and detailed discussion of the general question of consciousness and quantum theory.

Suggestions such as these are highly controversial. Many unscientific and overzealous speculations regarding the relation of consciousness to quantum physics have appeared in various non-scientific places, and have been severely criticized by some scientists[6]. I feel, however, that there is an open door to these possibilities as described below, and that conscious autonomy requires something that quantum mechanics may be able to supply. As our best and deepest-reaching physical science, it is, after all, it seems to me, necessary to consider the possibility thoroughly. Previous scientific speculations regarding consciousness and quantum theory have been suggested by Penrose, Hameroff, Ramon-Moliner, and Walker[7].

Evan Walker in particular has given a highly insightful and deeper reaching view of possible specific foundations for consciousness and its autonomous action in quantum theory. Walker sees partial clarification of the mind-body, autonomy mystery in its wedding with the measurement-uncertainty obscurity in physics. He takes consciousness as a directly known independent realm which relates to the physical brain as the 'observer' in the measurement conundrum. Through this consciousness is manifest as the Schroedinger wave function of some subset of fundamental particles of the brain, especially in the sense of its full range of possible states. Just as an observation of a physical system has been shown to unavoidably influence and thereby partially 'select' the state of that system, so consciousness is taken to partially select the physical state of the particle, and through this to influence the ongoing neurobiological processes of the brain. Walker further relates willful acts of consciousness to quantum mechanical phenomenon of 'tunneling', wherein particles mysteriously jump from one quantum state to another (for example, a change of position) sometimes in apparent defiance of energy and force requirements. By this association with 'observation'-related state change and 'tunneling', Walker's theory takes conscious autonomy outside the explicit determinative physical actions of the brain, thereby providing a possible resolution of the autonomy-determinism paradox.

Walker goes on to relegate consciousness explicitly to the wave functions of electrons acting in synaptic clefts, for here he finds enough energy in transmembrane potentials to tunnel electrons (quantum-mechanically) across the clef and thereby to alter configurational morphology of gating molecules so as to activate the synapse. He develops his theory to include coupling across multiple synapses, and, quantitatively with the help of certain critical numbers, to relate explicitly to neurobiological processes, and estimate quantities of interest, such as the relative strengths of 'willful' as opposed to fuller 'conscious streams' of data processing.

Thus, Walker very neatly interfaces consciousness with the deepest physicality of the brain as having access to a wide range of possible states for at least some of its important fundamental particles, and the ability to willfully alter some of these selectively. In this, the ultimate nature of consciousness is somewhat distanced from the brain's physicality and remains as we know it directly. The means whereby consciousness in fact produces the 'tunneling' state changes remains obscure, as they remain obscure in other observed physical systems. This remarkably insightful and provocative theory deserves the wide engagement, scrutiny, development, and possible modification by theorists of physics and neuroscience.

Walker's view leaves the nature of consciousness and its processes of selection outside of physics as 'what you see is what you get'. He also sees the energy utilized in tunneling as taken from the transmembrane electrical potential. These substantive qualities are hidden behind the veil of quantum mechanical uncertainty and the obscurities of the 'tunneling' phenomenon.

A second general view suggests the occurrence of a quantum mechcanical particle-carrier-charge complex as the undergirding of consciousness. This exotic possibility arises from the observation that consciousness seems to include some of the nature of a physical force. It is easy to imagine one's or another's consciousness as an entity with force-like qualities by its autonomous direction of one's body by means of its neurobiology. It seems possible to see consciousness in parallel with known physical forces of theoretical physics--perhaps as a force which both arises in and acts on particularly neurobiological physicality, as both gravity and electromagnetic forces originate in and act on dimensions of matter: gravity in and on mass, and electromagnetism in and on static and moving charges.

Might it be possible, for example, to find a force-particle complex with contextual properties like those of a conscious partial autonomy and particularly localizable to a certain subset of life forms? Something would be required to localize the effects to brain regions. Localization could be effected by association of the force with a certain type of charge (like mass with gravity or electric charges with electromagnetic forces). Further association of such charges within particular types of organelles or cells could serve to localize consciousness in the brain according to the localization of such organelles or cells.

Is it possible that we are so associated with a known physical force, or with a new or previously unrecognized force within the same general purview of current physics? Partial autonomy would be the major player in all this. Our autonomy, which the physical forces do not seem to have, might be different in this. Such considerations could help reveal the fundamental nature of consciousness and autonomy in relief against the known forces and elements of physical nature.

With regard to these possibilities, it is well to realize that physical theory is malleable as its gropes its way forward to account for the phenomena of the universe. It will surely try to accommodate itself to the experimentally established facts of both the microscopic and macroscopic universe. These developments very likely will retain the great bulk of structure of quantum mechanics, and its basic extensions into string theory. The site of action of consciousness and the forces of autonomy would be in the intermediate mesoscopic realm of biological or neurobiological life.

Such grand long-term efforts could come eventually to a fully satisfactory description of partially autonomous two-way interaction of consciousness and neurobiology and its apparent functional organicity within a foundational non-traditional physical framework. It is important to see that this would in no way diminish the mystery nor grandeur of either the universe or conscious experience as we know and cherish them. Alternatively, to disclude conscious partial autonomy with its particular difficulties, and then argue from the resulting structure that it doesn't exist because the remaining physical features seem well described doesn't seem cogent. It might be compared to having asked the neglect of the gravitational force because the standard model (before string theory) didn't contain a description of it.

6.4 Placing Conscious Autonomy in Physical Science

This section takes a fuller direct consideration of the mind-body question in terms of contemporary physical science as crystallized in the question, 'does the brain do it all?' There is widespread acceptance throughout neuroscience and much of the general public that this is indeed the case, although there is no satisfactory understanding of the substantive base of consciousness or free will. Some monists hold that autonomy is an illusion and consciousness an ephiphenomenon[8]. Statistical systems theorists point to the pervasive noise and uncertainty of brain activity as perhaps as open door for conscious autonomy[9]. Some see a quasi-, apparent autonomy in the unpredictable highly idiosyncratic activity of the individual human brain; others say the autonomy is real but obscure; most, perhaps, are simply open to these mysteries.

Alternatively, this book holds that: since consciousness exists, Ockham's razor implies that it must do something beyond what the physical brain itself would do without it; and that partial conscious autonomy is too clearly the central fact of human existence to be denied, and must exert some forceful influence on the brain's physical processes beyond what those physical processes would do in themselves. The alternative, that natural forces of the brain produce autonomous behavior is flawed because a conscious autonomous selection and action requires an independent integrative determinative freedom beyond that found in the traditional forces of nature. The only way the brain can be seen to 'do it all' is to make a substantive identification of consciousness with some non-traditional forceful entity in physical theory. The larger alternative, that consciousness and autonomy are illusions, is seen here as simply too much at odds with the intrinsic experience of every human. The only other option sees consciousness outside physics, yet somehow intervening into it.

It is precisely the possibility of identifying a non-traditional forceful entity in physical theory that makes both the statistical systems model and quantum mechanics central to the quest to find a suitable home for conscious autonomy in physical theory. Systems theory can point to the locations of 'noise' in the brain as candidates, while quantum mechanics may provide hints as to possible linkages between consciousness and neurobiology. Quantum mechanics includes both the phenomena of uncertainty and their description in a non-traditional mathematical physics.

What is needed are means of transmuting the non-traditional uncertain quantum events into macroscopic neural or neural-related effects and the identification of plausible physiological sites for such events. We can anticipate that these would be places where small numbers of microscopic atomic particles, such as electrons, might be placed so as to influence microphysiological processes. First, though, let's consider the nature of quantum physics a little more carefully.

6.4.1 *Interfacing quantum mechanical and classical physics*

Classical and quantum mechanical physics are two distinct ways of describing the dynamic behavior of any given piece of matter. Classical is better for larger size scales and quantum mechanics for smaller, namely the subatomic fundamental particles of matter. Quantum mechanics can be shown to predict, and so include, the classical rules for large numbers of particles and larger size scales. When the two approaches are applied to marginal cases where either might contribute, the quantum mechanics is the accurate view. Most theoretical physicists say that quantum mechanical effects are limited to the atomic and subatomic regions and that all larger collections of matter are safely in the macroscopic regions of classical physics. This is important not only for accurate scientific description, but also because of the vital distinction between the presumed exact determinism of macroscopic classical physics and the uncertainties of quantum physics.

Classical rules are found to apply safely down to at least nanometer scales, 10^{-9} meters. This is the current limit of mankind's ability to manipulate technological devices. Quantum effects take place in, and are generally restricted to, smaller dimensions of 10^{-10} meters and below. This value is the diameter of an atomic electron's orbit about its nucleus and is taken as the size of atoms. The nucleus, in turn, is about 10,000 times smaller at 10^{-14} meters, and its protons and neutrons and the quarks of which they are composed are at 10^{-15} meters[4].

The fundamental particles quantum mechanics consist of quarks, electron-like particles, and neutrino-like particles. The quarks reside in the nuclei of atoms, the electrons orbit these nuclei or float more freely in composite matter, and the other particles are more difficult to localize. Of the four forces of nature, the two non-classical forces (the strong and weak) are confined to operate on the microscopic scale because their strength diminishes rapidly over all but subatomic scales; they are the nuclear forces. The strong force holds quarks together in the nucleus. The weak force is responsible for certain particle transmutations in radioactive decay. The two classical forces (electromagnetic and gravitational) in striking contrast, have an unlimited range of influence.

A micron, one millionth (10^{-6}) of a meter, may be taken as a conservative estimate of the distance ranges of the neurobiology of neurons and the brain. This is well within the 'macroscopic region' of classical physics, 1000 times bigger than the nanometer level of human manipulation considered securely classical by physicists, and 10,000 times greater than an electron orbit of 10^{-10} meters and some some 10^8 and 10^9 times bigger than a nucleus and its quarks, respectively. The microscopic uncertainty ranges of quantum state variables are also normally exceedingly small.

Small Heisenbergian indeterminacies in quantum particle state variables (say position of atomic electron) are also very small, and should not be significant except at sites close to vulnerable microphysiological processes. These might occur, for example, in single atom-like particles on neuronal or glial membranes, synaptic interfaces, neurochemical or metabolic processes. More specifically, several possible mechanisms of forceful action have been suggested including: the release of single packets of synaptic transmitters by quantum mechanical fluctuations in molecular 'gating' at presynaptic neural membranes[10]; the quantum mechanical 'tunneling' of electrons across synaptic clefts as selected by the 'observer' in the quantum mechanical picture of the uncertainty of atomic measurements; the existence of a distinct quantum mechanical particle-carrier-charge complex for consciousness and conscious forceful action. Single organelles or molecules might transmute quantum mechanical variations into mesoscopic effects and establish localization within brain tissue. The glutamate receptors of the brain's glial cells have been implicated in consciousness by fMRI imaging techniques. The idea that conscious might be associated with the brain's energy supply (metabolism) through this is a powerful image (see Chapter 7).

It might be possible to translate such mechanism into conscious autonomous influences with an appropriate picture of the operations required by such conscious autonomous action. But first, it is instructive to consider the nature of quantum mechanical uncertainty a bit further.

6.4.2 *The nature of quantum mechanical uncertainty*

Quantum mechanical uncertainty manifests in two ways. The Heisenberg uncertainty principle states that particular pairs of particle state variables (e.g. position and momentum) cannot be specified simultaneously better than a specifiable limit. This is more than a restriction on observability. The simultaneous specification of both position and velocity for a quantum particle can be shown to produce results contradicted by experiment. The fundamental particles are in themselves in some indeterminable, smeared out, region about some given mean state. This uncertainty seems to be an inherent quality of the microscopic fabric of the physical universe itself Greene[5].

Further, yet consistent with this, all descriptions of quantum mechanical particles are in terms of probability density functions of state variables rather than specific numerical values. The Schroedinger wave equation, however, predicts deterministic evolution of these probability functions over periods between observations which disrupt the system and require updated specification of initial conditions.

The interesting thing about quantum mechanics is that it allows a possibility of physically non-determined selectivity and at the same time, an overall internally-driven determinative direction. Variations are generally very small but can be significant. They are inscrutable but lawfully behaved except when explicitly interfered with by an external observer after which new 'initial' conditions must be prescribed to reset the predictive description.

6.4.3 *Overall operation of conscious autonomy*

Conscious autonomy requires that consciousness sense and probably participate in an integrated or collective, often global, state of affairs, and then select and activate a number of appropriate target sites. Figure 6 illustrates this view and suggests that consciousness must:

• sense/apprehend overall state condition indicated by brain and body, (perhaps by systems of flags,see chaper seven, or globally or both),
• attain some overall sense of any necessary direction, intention, or adjustments required by these (abetted by deep integrative processes, conscious or not) , and
• manifest some kind of sensible selection and triggering of phsyiological targets to adjust/direct neurobiological activity accordingly.

6.4.4 *Types and principles of possible quantum mechanical effects in brain*

We can now try to pull together a broad view of the possibility of localizing consciousness in the brain. A primary principle adopted earlier in this work is that the placing of conscious autonomy in particular portions of the brain implies there must be some forceful, energetic interaction between it and physical neurobiology because otherwise consciousness is a deluded rider and autonomy an illusion. This need for autonomous forceful action, in turn implies that there must be some special physical qualities beyond those of classical physics in those portions associated with autonomous action because forceful interactions of consciousness with matter require it in order to resolve this with our descriptions of physics, and because it is necessary for localization of consciousness-- why, otherwise, isn't consciousness also elsewhere in the brain and beyond? This is also a fundamental underwriting principle of this work. In short, this principle implies that autonomy can't 'just be in' some portion of the brain without both non-traditional mechanisms and their physical undergirdings.

The following five processes seem the likeliest candidates for such quantum mechanical effects on the brain's neurobiology:

• direct molecular transmutations of quantum mechanical effects to macroscopic neurobiological effects. An example might be a quantum mechanical effect on the mechanical properties of one or more electrons in a gating molecule of synaptic transmission, resulting in alterations in the numbers of packets of transmitter molecules released and showing up as alterations in synaptic electrical currents and post-synaptic neuronal electrical potentials.

• similar transmutation in electrons involved in glutamate actions in the brain's glial cells ultimately manifesting perhaps in adjusted distributions of blood and metabolism to neurons in turn constituting an overall adjustment of the direction of neural activity (see Chapter 7).

• a general transmutational molecule or organelle to transmute quantum effects to macroscopic neurobiological or other biological effects generally. The existence of such an organelle could be a generic means for transmuting microscopic quantum mechanical effects, such as electron position, to macroscopic molecular or higher macroscopic levels of effectabilty which could, in principle, be applied in the same or different forms in different locations. The distribution of the organelle throughout the tissue of the brain would correspond to the distribution of the capacity for consciousness. Hameroff earlier offered an idea along these lines[7].

• an action of consciousness acting, as the 'observer' of quantum mechanical theory, on the brain's neurobiology or other biology, to 'tunnel' microscopic state variables through energy barriers (say, electrons across synaptic clefts) as suggested by Walker. This action is seen to operate as an observer-produced disturbance of intitial conditions within their Heisenbergian uncertainty ranges, between

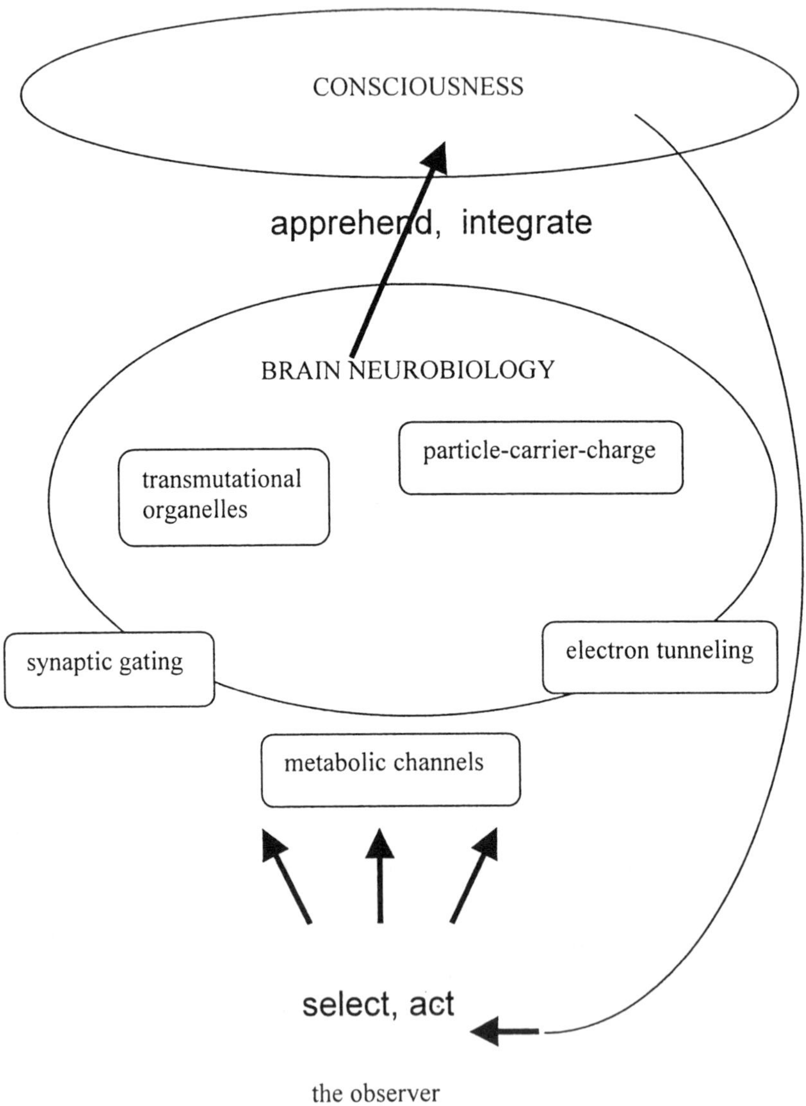

Fig. 6.1 Conscious autonomy and possible quantum effects.

periods described by Schroedinger's governing equation. The action is it at the palings, at intersections between consciousness (as 'the observer') and neurobiology, included as an boundary condition, in itself allowable by physics, but not in itself within physical law. Yet, Walker estimates its unique plausibility with candidate quantum mechanical effects by its compatibility with the energy level of the cellular electrical transmembane potential. This kind of action could be effected in other brain regions, for example in glutamate receptors of glia cells perhaps in relation to brain metabolism.

• entities and forces are represented in quantum mechanics by particles, force-carrier particles, and various kinds of 'charges' for the forces (for example, mass for gravity and electrical charge for electromagnetism). In physical theory, gravity and electromagnetism both originate only within their charges and act only on their charges. Perhaps there is a particle-carrier charge complex associated especially with consciousness in the substantive sense that consciousness is taken as one with these. This view would explicitly codify consciousness and its forceful actions and localizations within matter. This view also might more powerfully describe the sensing and apprehensions of consciousness according to a presumed localization of its charge within the matter underlying the apprehended qualities. All the previous views have spoken only to possible mechanisms of action.

In all the above views, the feature of autonomy would have to be preserved within the hypothesized operations of the consciousness portions of the theory. Any specific substantive identity material operational grounding of consciousness (such as with an organelle and its processes or a particle-carrier-charge complex) might be permitted to have lawful rules; yet, since consciousness and its autonomy would be seen as one with these processes, consciousness might seem to require some ultimate freedom of choice above even these.

This path of investigation could hope to illuminate the range of possible manifestations of consciousness in other creatures and the viability of other more exotic views of consciousness in biological and physical nature. It is clearly a difficult and obscure quest, but a suggestive path, and a highly important project which impinges on the ground stuff of our very nature and our relation to the world and universe around us.

6.5 Concluding Assessment of Quantum Mechanics and Consciousness

This work has been led into this thinking by the inherent contradiction between an existentially felt autonomy of conscious experience and its apparent identification with a brain whose physicality can do it all. This is more than a logical distinction; it resides in the comparison of operational qualities of two distinct realms of which one is thought to map into the other; its contradiction rests on the presumed autonomy of consciousness and on the simple existence of consciousness in itself. It might be resolvable within the physics of brain activity if there could be a way to identify an autonomy within the brain's physical operation and see a useful property of consciousness in such a view. For example, one might make a substantive identification of consciousness with some attribute of the physical brain, or, relatedly, suggest that consciousness, with its self-contained autonomy, arises as a transcendent quality in or around the physical-- analogous, say, to electromagnetic fields around the brain's neuroelectric currents, but somehow endowed with a freedom to select.

Selection is a pervasive quality of neural systems, and freedom of selection is an autonomy which is the critical quality of an operative consciousness. These must be satisfactorily characterized in any acceptable view of the consciousness-brain question. To illustrate the central importance of this point, consider the plausible minimal view of the brain as an indeterminate physical system which produces, according to its inherent internal neurobiological process, all the activities of the inner sensibilities, including especially all higher-order integrations relating to plans and behavior, reserving within its consciousness only the ability to select among these for ready action or for deferral with or without prodding for further development-- fundamentally an overall governing 'stop-wait, go' operator. In this simplified model view, the brain's natural physical integrative processes are seen as doing all the generation of candidate behavior patterns, almost all the selection among them, and the actions selected. Consciousness simply allows or disallows. In principle, one could imagine that such a minimal overall governing system and its allowing or disallowing could operate within some small region of indeterminism within a physical brain.

This simple model could be a drawing out of the possibility of indeterminism suggested by the statistical systems theory of brain function. Yet, and this is the central point, this leaves us with an important open question regarding the actual selection made within the variability available. If this variability is to house autonomous action, that action must be ultimately purposeful and directed: how, then, might such a consciousness actually sense, select, and effect its presumed allowing and disallowing? Mere specification of uncertainty or variability is not enough. It is not enough to say that consciousness is 'simply part of the system' without coming to terms with the necessary non-traditional mechanisms of action implied by conscious autonomy. Also, if it is all done by neurobiology, why is there consciousness at all?; and, in any case, how does it differ from or relate to the physical itself, and what is its nature? Note also that such a 'stop-wait, go' or other minimal consciousness could be supplied by extra-neural agents such as the brain's metabolic system by selective channeling of the blood supply to appropriate neurobiological locations.

The several possible types of quantum mechanical actions identified just above lead to differing conclusions. The identification of transmutation of small microscopic selections in the triggering of pre-synaptic gating of neurotransmitters to mesoscopic neuroelectrical currents provides a significant specific selected effect. Again, such a transmutation could be attributed to other systems like the brain's metabolism. Yet, the act of selecting itself in these cases is left open and so is the placement and localization of consciousness. These possible mechanisms, then, are incomplete in themselves and need to be characterized within a larger overall framework.

The idea of a generic organelle which transmutes microscopic quantum effects to mesoscopic neural results could satisfy the localization question yet remains open to the questions of autonomous selection and larger nature of consciousness itself.

The observer theory of Walker and the particle-carrier-charge hypothesis are more comprehensive views. They both absorb the selection process satisfactorily within larger frameworks, and the latter includes a means for localization as well.

Walker's observer theory puts the obscurities of selective conscious autonomy in the obscurities of quantum mechanical selection of particles for tunneling (synaptic electrons for tunneling across synaptic clefts) at the palings of physical theory and the physical brain, and beyond its governing laws within the veiled operations of the free-willed conscious observer who remains beyond our description. This placement provides the grandest, broadest, and most completely self-contained view in that it maps our physical neurobiological brain with its sentient existential consciousness onto our best current physical theory with its veiled observer. It is fundamentally one with the view that consciousness is ultimately outside of physics, yet identifies a locus where conscious autonomous effects may enter the brain's neurobiology. It maps brain with physics, conscious with observer, and conscious autonomy with observational disruption, tunneling, and 'initial' conditions.

Walker's view leaves the nature of consciousness and its processes of apprehension and selection in obscurity, and does not address localization. Yet it may be ultimately the best terminus, nonetheless.

The hypothetical particle-carrier-charge complex puts consciousness more centrally within physical theory by proclaiming a substantive identity of a physical (particle-carrier-charge) and an experiential (consciousness) face. Its physical state variables are subject to quantum mechanical indeterminacies, thus leaving an open door for possible outside adjustments, as are all quantum mechanical particles. This consciousness complex, however, in unique distinction from all other matter, also contains a third face, which is that of a functional autonomy consisting of the capacities for purposeful directive selection and activation of effecters, and some (minimal) integrative apprehension to inform these. The apprehensive and active processes are keyed to its charge which also selectively localizes them in the brain. In this hypothesis, consciousness could be seen at any moment as associated with some coordinated 'activated' subset of all those elements which contain the consciousness charge. All its processes of apprehension, selection, and target activation could involve adjustments of microscopic state variables all within the quantum mechanical indeterminacy limits. The microscopic effects could be transmuted to mesoscopic physiological levels as indicated in the last section, some of which might be in a special type of organelle containing the charge.

This is only a rough incomplete description, but sufficiently suggestive, I think, to indicate that it could possibly be developed to provide a plausible terminus for the consciousness-brain question.

This hypothesis, although plausible and suggestive, is, as all the ideas given in this section, highly speculative, uncertain, and requires a great deal more elaboration and study. For example, some characterization of its forceful actions and relations to physical laws is needed. The basic dynamic requirement of this particle-carrier-charge hypothesis is that its central processes of integrative apprehension, selection, and effecter activation remain autonomous and self-determinatively purposeful beyond any physical laws including those associated with the carrier force itself. In this, the problem area seems to differ markedly from those which have been so successfully resolved by the elegant determinative mathematical descriptions of physical laws governing electromagnetism by Maxwell, gravitation by Newton and Einstein, and quantum dynamics by Schoedinger. A completely different, and perhaps much simpler description may suffice.

The particle-carrier-charge hypothesis goes past the open door of indeterminism in itself in providing suggestive indications regarding biological mechanisms of conscious action and the nature of consciousness itself. It goes potentially, for better or worse, past the veiled obscurities of the observer theory in bringing consciousness into rather than at the outer margins of physical theory and locating it within particular physical properties of brain tissue. However, the fundamental question 'why is there consciousness?' remains illusive. We can only assume that it has to do with the capacities of autonomy. Perhaps further characterizations of the force and relations with physical law might illuminate this. In the meantime, and perhaps ultimately, the mystery of consciousness and its disjunctive relation with the physical, remains in all of these theories and in the classic mind-body question shrouded in the same obscurity as in the existential conscious experience of every human. It is interesting in this regard that both the particle and observer views of consciousness are themselves products of what they try to understand--our brains and the consciousness which tries to understand it all. So, this is an appropriate time to look a little further into the nature of this individual conscious experience in itself, which we will do in the next chapter.

First though, notice that this entire approach, and especially these last two observer and particle-carrier-charge examples, identifies different shades of intermediate ground between the traditional dichotomy of hard monism and dualism in the brain-mind question.

6.5.1 *Does the brain do it all?*

It seems the brain needs consciousness and autonomy to do it all, and that the place of these in relation to the physical brain is not clear. The question is whether these are placeable within physical theory or stand outside of the physical in some vague or specifiable sense. They might be localizable in quantum mechanical terms as the possibilities suggested above, and they might be associated as much with the glial cells and the brain's energization as outlined in Chapter 7. They might stand outside physical science in as the quantum mechanical 'observer' as Walker suggests.

6.6 What Are We? Fundamental Views of the Nature of Consciousness

In any instance one might say 'what are we?: well, it seems we are some sort of original source of causal energization and experience associated particularly with our own body and especially in two-way mutual interaction with or within our neurobiology, and through these with the outside world. In this we may see ourselves (conscious experience) as superordinate partial director of the highly complex global neurobiological functioning of the brain and nervous system. We ride along, experiencing, reflecting, and directing with or upon this neurobiology, and therefore with and within the further underlying biology and physicality of our body, and through this body interacting with external reality, representations of which we accumulate and reflect upon. We receive energetic flux from external reality, much of which is of independent origin and some of which reflects back from our own energetic projections.

All this may eventually be included in some appropriate expansion or application of physical theory such as the transmutation of quantum mechanical effects indicated above, or it may simply stand outside of theoretical physics. Alternatively, it may be that monists will find an alternative convincing interpretation of partial autonomy within traditional physics and neurobiology--but I doubt it. The common view of consciousness (outlined further in the first paragraphs of the next chapter) indicates several concerns regarding some aspects of the physically-based scientific view. Individuals feel a deep protective identity as one with their conscious awareness, their experiences therein, and their various conscious capacities, including free will. The idea that the brain can do it all with its neurobiology alone seems to contradict one's autonomy and thereby fly in the face with one's most vital and immediately sensed capacities. Further, it seems that conscious experience is meaningless without autonomy: If the physical brain can do it all, then why is there consciousness at all? This direct feeling seems linked to a logical contradiction within science itself: If a scientific theory of a human brain-mind were being constructed from the outside, as it were, intending to ground behavior in brain processes, then Okham's razor (the principle of the least sufficient hypothesis) would disallow the assumption of consciousness. Since consciousness is a fact, it would seem that it must do something that the brain itself does not (or do it better which is the same thing), and thereby be somehow beyond the brain. Such concerns center around the basic question 'can the brain do it all?' We will now suggest five main responses, two grounded in brain science, two in the existential view, and one grounded in both.

The basic hard monist position is that all of consciousness is determined by determinative physical operations of brain (and perhaps some of body), and autonomy is only illusory. Consciousness is a deluded rider in a slushy neurochemical machine. I think this position asks us to give too much away.

A slightly softened monism suggests that the rich complexities, unavoidable random uncertainties, and imperfections of brain operations, supportive body conditions, and ambient environments, and the vast amount of idiosyncratic conditioning and learning of any individual produce a quasi-autonomy which is, for all practical purposes, indistinguishable from a real autonomy in the sense that one makes choices based on one's unique individuality which are not random yet are in practice unpredictable[11]. Many stochastic system theorists, for example, see the brain as inherently non-deterministic because of its pervasive and obscure 'noise'[9], and some hold further skepticism regarding traditional physical determinism on larger philosophical grounds[10]. I see this view of consciousness as flawed at least by the 'deluded rider' syndrome: our sense of autonomy is not authentic if it simply reflects neural processes, be they deterministic or not. I believe, further, that in the absence of quantum mechanical uncertainties, the brain's neurobiology is essentially at least strongly deterministic, if not exactly so in the sense of traditional physics, even if we are unable to measure or state all the quantities necessary to calculate its course.

The third intermediate possibility may seem an exotic one, but in fact is a conservative scientific suggestion which integrates consciousness within physical science and within the brain. This view suggests the possibility that consciousness is a fundamental component entity in its own right which might be characterizable as a fundamental force-'matter' complex according to the rules of quantum mechanics fashioned now through string theory. Consciousness, by virtue of its universally-sensed partial autonomy, would entail a force-carrier particle and represent a newly-recognized fundamental force (to join the strong, electromagnetic, weak, and gravitational forces). By the super-symmetry principle the force particle would be accompanied with a matter particle or particles. There would be some kind of 'charge' associated with the consciousness force. Perhaps such particles could be associated with some existing or new atoms, which in turn might be in molecules or organelles associated with particular cells or volumes of the brain or its associated body support systems. If such consciousness force particles eventually act on neurobiological processes which are mediated by traditional matter (ie, not containing consciousness particles) then this 'receptive' traditional matter would need to have some non-zero 'charge' for the consciousness force, just as ions are charged for electrical forces. We see this as a plausible possibility which would resolve the compelling claims of conscious experience with the scientific quest to reliably ground consciousness in the physical processes of the brain.

A fourth view is that consciousness is a fundamental field in its own right which exists in intimate interrelation with the brain, and that we may or may not eventually come to know it better. This possibility is a less explicit form of the third, and a more open statement of the fifth. The theory of Walker[7] outlined above is an excellent example of this class. Popper and colleagues also indicate a model of this type[12].

The last possible view is that consciousness exists as a fundamental field in its own right and is not only obscure but ultimately mysterious, beyond our vision. Here, we are our consciousness and autonomy, and can or already know it by introspection as well as it can be known (except finding its detailed interactions in body and brain); consciousness is a mystery like the origin and contexts of the universe and of life in molecular biology and otherwise, and with them, forever ultimately unknowable to us. This is an hypothesis of obscurity, a view that 'what you see is what you get'. Well, perhaps, but let's keep looking anyway.

This discussion is continued in the next chapter, turning to an overall interpretative description of the existential phenomena of consciousness and a functional theory of its operations within the brain.

Endnotes

1. Edwards1967, Flanagan and Kim1995
2. Davies1988, 1990, 1991
3. Ridley1999, Keller2000, Tudge2000, Watson2004
4. Pollack2003
5. Greene1999
6. Stenger1995, Wegner2003
7. Penrose1989, 1994; Ramon-Moliner1994; Hameroff1994, Walker2000
8. Wegner2003
9. Korn and Faure2003
10. Lewis et al 1997; Lewis, personal communication2005
11. Kercel et al2003
12. Popper et al1993

Chapter 7

A Functional Theory of Consciousness and its Relations in Brain[1]

The study of consciousness inherently revolves around two central observations associated, respectively, with its two poles, consciousness and the brain. The first is the immediate universal sense of any human that he is one with his sense of experiential awareness and autonomously directs the living of his life from this center of awareness. The second is the strong association and dependence of conscious phenomena on the brain. A fundamental theory of the subject must answer to both these perspectives.

From the first perspective of consciousness itself, a rather widespread common universal human view of human existence emerges which emphasizes: especially, the seemingly free ongoing guidance and direction of one's life; conscious experience itself; the direct sensory experience of one's environments and of recollected memories; reflections and thinking. One comes to a sense of consciousness and vitality in oneself and in others as a life-directing almost tangible quality of spirit, deeply significant and markedly different from the harder physicality of the ambient world; to existential involvement with feelings, possibilities, potential, values, meanings, peak experiences, obscurities and mysteries of existence, and ultimately with senses of wonder and wary partial kindredness with other deep energizations of the universe, world, and their things.

Science, on the other hand, eschews much of this common view and seeks to ground consciousness (and life) in the physical universe, looking to an eventual harmonious continuum through physics, biology, and psychology. These three realms of science seem to reflect three nested strata--the physical, the biological, and the conscious brain. Each stratum has its own self-relevant patterns of activity, organization, and dynamics. Each stratum has its own obscurities of origin and essential nature. These obscurities define two central questions in the grand synthesis of science, namely the presumed origin of the DNA molecule within the physical forces of nature, and the essential nature of consciousness with its linkages in the brain.

This chapter presents a functional theory of consciousness based on a view of its evolutionary origin and an introspective view of its fundamental phenomena which may be readily approached by any human. It discusses these foundations, the theory, and the localization of consciousness in the brain's neurobiology.

7.1 A Functional Theory of Consciousness

7.1.1 *The flowering of consciousness*

The fundamental hypothesis taken here is that neural plasticity, self-direction, consciousness, and autonomy, although all existing to some lesser extents in earlier forms, underwent an explosive collective evolutionary flowering in humans which defines our essential existential and functional nature. Specifically, this has created what we can see as a fourth level of hierarchical higher central control over three prior nested levels[2]: autonomic and brain-stem responsiveness (homeostasis, startle, arousal, attention, and fight or flight mechanisms); instinct (with its 'fixed action patterns' and 'trigger stimuli'); mammalian processes of appetitive behavior (with particular sensitivity to 'incentive stimuli' and 'action-contingency situations') and affective behavior (introducing some higher levels of significance or value-weighting).

In humans expanded plasticity and consciousness give rise to greatly expanded self-correcting plastic adaptivity (with extensive plastic inner constructions) and to higher levels of detached reflective valuations, judgments, and overall governance (with further extensive plastic inner constructions).

7.1.2 *The phenomena of consciousness*

It is notable that, of all the phenomena of the brain, the surface level phenomena of consciousness are, by their very nature, the most open to common introspection. These as can be readily classed according to two dimensions: as associated with either external here-now experience and activity, or with internal representations or states; and as either passive receptions or perceptions, or consciously-directed active processes. Further, consciousness has a pervasive involvement with vast arrays of learned and constructed internal plastic patterns of arrangement many of which are used to govern our understandings or responses to the things of our ambient worlds, or to our behavior in them. These can be labeled as 'inner constructions', useful or not. Many may be seen as 'adaptive constructions'; others may be more freely conceived as free creative constructions. Yet further, conscious-directed activity relies heavily on mechanisms of attention and of searching or questing, including especially thinking, arranging, and planning.

Passive receptions of the here-now include sensations and images of sensory bombardment, which are among the clearest of conscious experiences. Passive inner responses include sensations of mood and feelings (sorrow, joy, anxiety, hostility, ...) and various 'flags' of significant inner judgments or recognitions (for example, the 'aha' phenomenon of a subconsciously completed solution of a mental problem, or the similar inner revelations of unsought judgments, for good or ill, of things or persons in one's life). Passive inner conscious experience also involves the immediate sense of conscious awareness in itself, including a sense of 'conscious space', and various rare but significant 'peak experiences' including deep mystical experiences. Passive inner experience includes the partial marginally conscious phenomena of dreaming and its subconscious dimensions.

The active behavior of humans in here-now living is driven by mostly conscious overall guidance which intermeshes the large array of inner constructions (particularly a large subset which we can label 'script-plans-intentions') with the interpretation of ongoing sensory bombardments within other interpretative inner construction patterns appropriate to the overall context. In humans a very large amount of active conscious experience involves consciously-directed cultivation of the inner constructions by active thinking or reflecting as driven by quest or search processes utilizing an attention mechanism.

7.1.3 *Fundamental tenets of the theory*

Consciousness blossomed in a joint explosive expansion with higher neural plasticity, self-direction, and autonomy.

Its main surface-level manifestations are: conscious awareness itself; sensations and images of sensory perception; reception, construction-modulation, and utilization of extensive organizational neural structures (inner constructions); partially autonomous overall instigation and guidance of action.

Conscious awareness is in part like an almost empty space into which particular elements from almost any part of the nervous system may variably enter--either as temporary standing residues; or more particularly as in externally-driven sensory bombardment or consciously-driven foci of attention; or as in subconscious inner generation (flags).

This work sees consciousness as the central component of a small but vitally significant overall existential-functional control system which senses and acts globally over nearly all of the nervous system so as to exert central control of the ongoing behavior of the human organism--the highest level of control in humans. This integrated conscious system contains all of consciousness itself and also some extent of associated subconscious processes, and all the structures directly associated with both.

The consciousness system operates on the simple basis of responding to the overall configuration of flags and other measures of informational, behavioral, neurobiological, and biological states with a corresponding redistribution of internal activations or energizations.

In this, the consciousness system would not necessarily be in itself very 'smart'. It could simply energize informational and other smart subsystems appropriately so as to fine-tune the adjustments in the direction suggested by the initially observed global distribution of significant flags, by modulations directed ultimately, either internally or externally as these smarter systems would also determine. Minimally, the only intelligence of the consciousness system would be where to redistribute the activations. The activations themselves could be general generic. Further, they could be either neurobiological (neuroelectric or neurochemical) or, for example, metabolic (see below). Alternatively, the consciousness system may easily be taken to include more smart processing without altering the structure of this theory. (For example, it might include some relatively small informational networks to represent some degree of particularities and operations regarding the nature and distribution of initial global state and flags, and of the resulting distribution field, and the transition matrices between them.)

At the grandest level, the overall control embodied in the operation of the consciousness system (seen here as effecting a redistribution of global activation-energization according to a global system of focal flags) may be imagined to follow some kind of guiding principle relative to these distributions. This could be perhaps cast as some sort of minimization principle in a generalized reflection of physical science. The minimization of cognitive dissonance[3] could possibly be the manifestation of this principle in application to cognitive informational flags. Much of the detailed operations of this overall guiding principle might be worked in the smarter systems activated by the consciousness system, likely in their neural networks. Flags themselves represent any item that has been revealed by the nervous system to need attention. The nervous system represents these as temporary standing patterns of neurobiological activity, say an active recurrent neuroelectric pattern or a neurochemical accumulation or imbalance.

This work takes the consciousness system as operating from a region around the outer edges of the brain's informational systems but often in close interrelation with these, as it seems humor, creativity and much of questing thinking seem to do. This allows us to function without excessive encumbering content, a constructive and autonomous place beyond potentially erroneous or inadequate content, and, perhaps the most intriguing, an enhanced potential to behave reasonably well in situations we don't understand.

Consciousness is taken as somehow embedded in some parts or

processes of the brain (including its support systems), and as normally operating with intensive interactions with particularly the memory banks and inner constructions of informational networks, and also other systems, but as functionally and perhaps fundamentally distinct from all these other higher informational and related networks.

Many brain processes are subconsciously active and only sometimes conscious. The processes and elements of memory banks and inner constructions, including the informational system, for example, are taken as subconscious--outside of consciousness--except when consciousness actively focuses questing attention on some informational region, be it an 'inner construction' or otherwise. (Relatedly, introspection on active constructive thinking, for example, can reveal that often the first stages of solution-recognition in constructive thought precede the revelation of the substance of the resolving thought connections. These latter are often consciously revealed only in a secondary wave of attention and then mostly with the explicit assistance of language.)

Informational and other neural networks of the brain are capable of doing a great deal of self-organization and self-direction in themselves subconsciously without any involvement of consciousness. Consciousness is integratively cooperative with these systems and makes extensive use of them. It greatly increases the effectiveness of these and more importantly quests these by thinking, thereby building a greatly expanded range of internal representations, valuations, and reflections (truth, aesthetics, wide range of values and value-weighted cognitions); and adds a greatly enhanced richness to human existence and apprehension; and greatly enhances the direction of behavior and the cultivation of internal constructions.

The possible localization of highest conscious control in metabolism would supply several advantages. Control would be more easily shielded from entrapment by irrelevant neuroelectric energizations; control could more easily 'push away' by redistributing underlying metabolic energization. It could avoid encumbrance by irrelevant content, and more flexibly redirect activations according to the more significant ones. Most fundamentally, it would put central control close to the ultimate source of power of the brain and to its very life itself. This would seem the most secure bodily strategy.

7.2 Characteristics of The Theory

Consciousness seems sometimes like a space in which separable elements of experience may come and go, and sometimes more like a flashlight which illuminates various realms in association with attention.

Attention has both variability of size or scope and variability of depth. For example, one might quest for and find whether or not there was anything extraordinary within a loosely related set of encounters and then be done with it, or quest further for particulars within one of those. The degree of recollected information illuminated by conscious attention is considerably different in both breadth and depth in these two instances.

Epileptic seizures and highly threatening moments are striking examples of highly intense conscious experience. Seizures entail painfully high degrees of awareness across wide ranges of content; threatening moments entail intense awareness of highly localized content, to the extent that even the most rudimentary normal actions are neglected or muddled (even those normally performed with little conscious attention). Normally, total overall intensity or energization of conscious attention seems to be kept within a normal range, either broad and shallower or focused and brighter. Perhaps epilepsy is fundamentally related to a malfunction of a control within the conscious attention mechanism. This would predict a tendency for higher and more variable intensities of attention and mental focus in epileptics. Heightened levels of consciousness of less striking intensity also frequently occur in such higher-level behaviors as play, adventure, competitive sports, certain kinds of interpersonal encounters, and a range of peak experiences.

A number of significant human phenomena seem to take place around the outside edges of cognitive informational banks. These include humor, which often seems to be associated with the failure or convoluting of precisely our conventions and inner constructions of behavior or understanding. Creativity is also such an edge phenomenon. Human inner creations often reveal themselves as a felt presence emerging from hidden and obscure regions of the mind, normally following a period of intense focus and struggle with a relevant quest.

As indicated above, deeper integrative conscious human thinking also seems to operate in initial stages with associations and connections beneath or outside the realm of the consciously perceived seeds of the quest.

The importance of dreaming to human mental state, and its occurrence in the absence of normal consciousness strongly suggests the existence of deep subconscious integrational operations within informational and related systems. Dreaming also underscores the distinguishability of normal consciousness from informational banks and their ongoing processes. It is as if two complementary systems are involved in the ongoing construction and maintenance of our inner informational banks and inner constructions, which is just as we suppose here. Further, the vague masked consciousness associated with dreams may hint at an enhancement of this integration by the consciousness system.

Although will and partial autonomy are vital properties of consciousness, nonetheless the deeper reaches of consciousness seem beyond volition, where these have been given over in a fullness of open awareness in itself.

The following speculative suggestion might indicate a pressure or tendency or drift of central control to ever more peripheral regions at the outer limits of preceding levels, such as: An early level of representational integrations in neural networks; expansions therein to higher levels of abstraction; a shift to metabolic control. The emergence of the consciousness system itself may be one of these processes or separate, prior or later. Perhaps a shift to the adoption of a still unknown existing or new force-matter particle complex in brain tissue is involved at some stage (see Chapter 6).

The larger principles of functional organization given in the preceding section imply a wealth of brain engineering. The use and cultivation of the inner constructions by the consciousness system, for example, indirectly engage the operations of a number of neural mechanisms mostly through the intervening directions of various informational, affective and feeling, valuative, and other systems. These include: direction of attention, quest or search, selection, learning, recognition, recall, appraisal, valuation, judgments, choosing, release, activation, anticipations, planning, selection and initiation and guiding of actions, and so on.

7.2.1 *Summary of the processes of this theory of consciousness*

This chapter has suggested that consciousness is the central member of functional system of the brain which serves primarily to help guide individual living through intimate cultivation and use of inner constructions. It may operate most grandly as a high-level feedback system which redistributes energization-activation according to a measure of global well-being as represented in and by the brain. It uses flags and energization-activations, and has autonomy. Its main functions relate to: awareness itself; global well being; here-now living; and inner constructions. These in turn imply a host of contributory attributes including: {threats, emotions, feelings, moods, intuitions; incentive stimuli, action-contingencies; sensory images and sensations; genetic prescribed instincts, fixed action patterns, reflexes; intrinsic, spontaneous, learned, and willful actions; learning-memory, significance-recognition, value-weighting, inner constructions; thinking, plans, intentions.}

7.3 Localization of Consciousness in the Brain

The localization of consciousness in the brain remains obscure because of its variable content, uncertain range of operations, the undesirability of doing invasive experimentation on human brains, and its apparent trans-physical texture. We might imagine consciousness as associated particularly with some particular region(s), function(s), process(es), structure(s), or mechanism(s) of the brain. These might be neurobiological or metabolic, or some other brain function or energization (Cannon). Metabolic structures, composition of extracellular space (local, regional, global), neural networks (local, regional, global), certain cell type(s), organelle(s), ion(s)-transmitter(s)-molecule(s), even perhaps atoms or fundamental particle(s) of nature might be central. Most likely, consciousness might be multiply grounded, and make use of several of these possibilities and perhaps others. We will here identify a number of suggestive processes as possible partial correlates of consciousness, in two groups and comment on these:

7.3.1 *Possible neurobiological correlates*

Phylogenetically, the natural occurrence of higher control of brain activity can be seen in four major levels of functional organization[4], progressively nested in a hierarchy which centers highest control respectively, in: brainstem-midbrain, midbrain-limbic-cortical, limbic-cortical, and consciousness system. Autonomic, startle, arousal, and instinct are representative of higher control in reptiles and fish with recognizable antecedents and parallels throughout the animal kingdom. These processes may be seen as largely spinal, brain stem, and primitive midbrain. Mammals add in higher governing processes of appetitive and affective behavior. These bring in additional operations of the midbrain and limbic system and rudimentary cerebral cortex. Humans add in yet higher governing processes of extensive plastic adaptivity, and freer and more richly informed overall conscious reflection, valuations, inner constructions, and choice. These include especially extensive involvement of the cerebral cortex and hippocampal system, and also extensive global interrelations with midbrain, limbic, brainstem and other regions.

In humans, virtually all large brain regions are implicated to some extent in various sometimes conscious processes. All the following have been so touted: the brain-stem (arousal, attention, fight or flight, regional emotional centers of reward, noxiousness); the limbic system (emotions, neurochemical transmitters of 'states'); the hippocampus (experiential learning, embedding of memories, value-weighting, pre-cortical structurings); neocortex (sensory perceptions, perceptual and logical memory bank, informational processing, inner constructions, language, planning [frontal]); substania nigra (will, also with basal ganglia and thalamo-cortical loops[5]). Further, both broader reaching global configurations and important processing engaging the midbrain may be involved as functional units.

Important processes which in themselves may associate more or less readily with consciousness include especially neuroelectric activity, manifesting either as: transmembrane signals in neurons (graded generator potentials or families of related dynamic patterns of spike trains in neurons or neural networks--sequential configurations or synfire chains); or as their associated local, regional, and conglomerate currents loops which traverse neuronal membranes through intra- and extracellular volume fluids in association with all neuronal electric signals, and the corresponding electric and electromagnetic fields which accompany these currents.

Coordinated collective firing patterns in systems of networks seem to be our best candidates for the images and sensations of direct sensory bombardment, and for the representation of higher order cognitive informational representation and processing in the brain (thoughts, reflections, inner constructions, memory banks, many kinds of representations and alarms)[6]. It seems likely that such firing patterns relate directly although variably with consciousness. The association of current loops and their associated electrical and electromagnetic fields with consciousness is a highly suggestive possibility, but leaves one disbelieving that extrinsic and man-made electrical or electromagnetic fields might be conscious, wondering how then to imagine that those in the brain should be so, and why some and not others.

Fluxes or pool of certain ions, regulators, or other molecules which govern neurolectric signaling or neurochemical process, or chemical states in themselves might relate directly to consciousness of feelings, moods, emotions-{synaptic transmitters, neuroregulators, hormones and fluxes of the endocrine system, chemical compositions of intracellular, local, regional, or volume pools[7]}. Poznanski has related consciousness to volume transmissions generally[8].

Perhaps particular cells contain some property or ingredient that facilitates consciousness[9]. For example: cells with dendritic trees that encourage confluence of extracellular currents (pyramidal cells, 'poker chip' cells of the reticular formation); a cell in itself, for example a pyramidal cell; a cell endowed with an organelle that has conscious-related qualities (Hameroff); electrons that have 'tunneled' (Walker).

Others have studied neurobiological foundations of consciousness from the standpoint of various psychological functions such as cognition or emotion more broadly across neurobiological foundations[10].

We can readily see that many of these possibilities may play a part in the full fabric of consciousness, and that consciousness may well have some direct experiential contact or operational interrelation with more than one of them. The flags which consciousness interprets as items needing attention seem likely to reflect certain temporary mostly local situation-dependent standing patterns of say neuroelectric activity or neurochemical accumulations or imbalances as generated by the integrative processes of the nervous system.

7.3.2 *Glial cells and brain metabolism*

Recent experimental and theoretical studies have implicated the glutamate receptors of brain glial cells in consciousness[11]. This is an extremely significant finding and is consonant with the energization view of consciousness presented here. In the present theory one could imagine a redistribution of metabolic energization directed to regions containing those sets of neurons and other neurobiology which could resolve the active flags. These neurons and all their projected associations would become more active neuroelectrically and neurochemically, thereby driving the resolution of the flags by the intrinsic interconnective integrational logic of the brain. The resulting change in the global distributions of flags would then redistribute the metabolic energizations away from regions of resolved flags and to others as necessary.

7.3.3 *Broad interpretations regarding localizations of consciousness*

The larger questions here are of the localizations and then autonomous actions of the volition and will. The images mentioned above have suggested the possible associations of consciousness particularly with elemental extracellular current fields or metabolic-related substrata, and perhaps particularly with higher regions of the nervous system like the brain stem recticular formation and lateral hypothalamus; the limbic, hippocampus and perihippocampal regions; and regions of the cerebral cortex. These can be seen as a large interrelated realm of tissue around the head of the brain stem. Further cellular possibilities of localization will be indicated below.

One might couple this with the experiential sense of an inner reflective space of conscious awareness. Autonomy is associated most deeply with this inner theater of conscious awareness itself, deeper within than any content of the theater. Indeed, the deeper regions of the theater are obtained by letting go of will and its attachments and giving over one's autonomy in a fullness of open awareness in itself. We might envision an association of this inner reflective space of conscious awareness with some region of the brain and label that region a 'central integrative space', imagined as inclusive of appropriate processes and neural tissue. Autonomy also would be associated with this central integrative space. One might tentatively associate this central integrative space with the region around the head of the brain stem indicated above.

Several important papers have considered the possible underlying unity or fragmentation of consciousness[12]. Poznanski sees a conscious state in humans as the result of an integrative neurobiology intimately linked with the language centers in the cerebral cortex (e.g. Broca's area and Wernicke's area), and motor-sensory areas in all other species. In his view this is governed by the dynamical pressures inherent in the chemistry and uniquely positioned within the architecture of the brain in an incoherent precision yet to be fully mapped. The incoherent dynamic process which unifies conscious states is called 'functional productivity' (Globus). It is an incoherent process because conscious states are subjective or first person ontology (Searle). Quale is also a term that would fit under the umbrella of the self-referential character of the brain explicitly manifested through functional productivity and dependent on the dynamics of the signal and the phenotypic adaptability . Poznanski points out that functional productivity supports Searle's unified consciousness theory (i.e. the production of any state of consciousness at all by the brain is the production of unified consciousness), and notes, however, that an incoherent process is in contrast with the views of Tononi and Edelman that both integration and differentiation (or segregation) co-exist, where the latter is the result of large clusters of neurons firing in a synchronized fashion.

7.4 Comments

Good theory anywhere in Science, and especially in the complex field of consciousness, needs to get to the heart of its subject matter--to recognize the central role of inner directive causes. It is a question of the organization of observations according to these inner causes and their resultant governing dynamics. A good theory and approach needs to provide a substantive operational framework within which things can be interpreted, not merely classed. Then it needs to spell things out. Popper has well described this view of the nature and role of good theory in a developing scientific area[13]. He sees that the most useful theories are often those that bring on their own supercession most quickly by addressing fertile obscurities at the margins or depths of a developing area.

In consciousness study, the heart of the matter, the inner directive cause, is autonomy and its directive governance of the guidance of behavior. It has become commonplace in the neuroscience of the last twenty years to speak of consciousness as an emergent quality of the nervous system[14], sometimes as if this label diminishes its essence. But consciousness is no secondary fact or mere quality of the nervous system; it is a vital central factor in the overall functional organization of the human brain and the vast majority of its inner constructions. Hypotheses regarding this should inform the broadest reaches of brain theory. Conscious autonomy also engages the deepest philosophical question of brain, its relation to human living, and its relation to the very nature of the universe itself. It suggests the need for fundamental expansions in the field.

Theories are most productive which: speak to the underlying directive causes in a view consonant with the deeper inherent constraints of its subject matter; provide a functional interpretative framework for its observable phenomena; relate to important neurobiological findings; provide testable predictions regarding both the existential and neurobiological faces of consciousness; suggest possible physical substrates for consciousness in the brain; provide interpretative ground to guide questing experimentation and invite such experimentation; and suggest fertile outer limits and invite their own adjustive improvement or supercession.

This chapter, the last of three focusing on consciousness, has outlined the central function of consciousness as the partially autonomous direction of behavioral living in accordance with vast systems of plastic patterns of inner construction, and the development of these--all in the service of a central concern with widely perceived well-being. We now turn to three chapters on human cognition. Chapter 8 describes language as a cognitive inner construction which mediates between consciousness and the neurobiological understructure of its various contents. Chapter 9 takes conscious awareness as the necessary first source and theater of our approach to first principles of human awareness, apprehension, and the knowing of things. Chapter 10 characterizes the relations of knowing with neurobiology.

Endnotes

1. A modified version of this chapter appears as a paper in the Journal of Integrative Neuroscience, Vol 3, No 3, 2004, by agreement with Imperial College Press.
2. see MacGregor1987, 1993 for reviews
3. Festinger1959
4. Sarnat and Netsky1974, Brodmann1909, MacLean1970, Luria1966, MacGregor1993.
5. Groves1983, Groves et al1978, Wilson et al1982
6. Gerstein et al1978, Abeles1982, Edelman1987, MacGregor1991, Poznanski2002
7. Cannon1944
8. Poznanski2001
9. Schiebel and Schiebel1958, Hameroff1994, Walker2000
10. Taddei-Ferretti and Musio1997, Kazniak1998
11. Magistretti and Pellerin1999, Poznanski2002, Searle2000
12. Searle2000; Tononi and Edelman1998; Poznanski2001, GBT, 2002; Globus 1992
13. Popper1959
14. Koch2004, Crick and Koch2003, Porter1994

Chapter 8

Consciousness, Inner Constructions, and Language

Chapter 7 has described consciousness as fundamentally involved with the construction, development, and use of large systems of inner constructions, many of which are more or less associated with the guidance of behavioral living as one has encountered one's ambient circumstances and may expect to so continue. We now turn to three chapters on human cognition. This chapter describes language as a cognitive inner construction which mediates between consciousness and the neurobiological understructure of its various contents. Chapter 9 takes conscious awareness as the necessary first source and theater of our approach to first principles of human awareness, apprehension, and the knowing of things. Chapter 10 characterizes the relations of knowing with neurobiology.

8.1 The Range and Nature of Inner Constructions

Most Inner constructions engage basic views of the properties and nature of the surrounding worlds; the people, creatures, and things in it and their ways; one's personal dangers, wants, preferences, and opportunities; one's personal ambient circumstances and current overall life plan; the larger and deeper meanings of life and purpose; and so on. Some important inner constructions consist of life scripts, plans, and intentions. Many are simple motor skills. Many are more abstract adaptive integrations of various levels. Others are freer, relating more closely to the higher sensibilities (aesthetics, understanding, meaning and value) in themselves. Others are simply playful or whimsical. Many have to do with local personal interrelationships. Many are bodies of learned knowledge, obtained formally in schools or in varied personal experience. Much reflects sporadic idiosyncratic experience.

Structurally, some inner constructions are intrinsic in themselves, such as the neurobiological patterns of instinct, or archetypical templates for the sensory recognition of certain significant forms or the like. Most, however, are plastic patterns the constructions of which require the active participation of consciousness. Inner constructions range from mundane to esoteric; personal to universal, objective, and transcendent. Many are plastic constructions built upon some base of intrinsic operational structure such as dancing upon basic locomotion. Inner constructions of basic motor skills such as playing the piano or basketball, are formulated and stored within the vast hyper-regular circuitry of the cerebellum which is devoted the cultivation of finely-controlled patterns of movement and action. Very many plastic inner constructions have to do with one's overall life purposes and codes of behavior (a script), or progressively more current plans and intentions. Objective and universal inner constructions are created and found in relation to the higher valuative sensibilities, aesthetics, apprehension-cognition and value. An important bank of inner construction is the experiential record of one's life-long consciously attended experience found in the perihippocampal region[1]. These memories seem to be recorded automatically simply by virtue of one's conscious attention, but automatically outside of one's conscious will in itself. Language is a widely and highly useful central system of such inner constructions.

8.2 Language

Language is a large plastically-learned, internally-interconnected system of symbols (words) which represent meanings with compatible spoken and written manifestations, which serves personal expression, internal cognitive mapping and understanding, and interpersonal communication. The symbols of language, words, can be taken to represent entities, actions, qualities, or relationships, from simple specific to highly abstract. The construction, use, and development of language, as many higher inner constructions, are mostly directed by consciousness. Language is mobile in the sense that conscious attention can direct conscious awareness to any given region of language. It is open-ended in that new elements (words) can be defined without limit. Language can be directed beyond itself into new realms. This can create increased discernments, increased scope of comprehensiveness, or higher dimensions of abstraction. Language can be sent on quests in pursuit of the questions: what?, which?, where?, when?, who?, how?, why?

8.3 The Structural Nature and Place of Language in Neurobiology

One may see language as a construction of the inner sensibilities to mediate between consciousness and the fuller neurobiological systems it needs to gauge and guide. In this language reduces a rich, complex, ambiguous, perhaps quasi-chaotic neurochemical, neuroelectric complexity to a discrete set of fixed unitary symbols (words).

8.3.1 *The outer conscious-related face of language*

Words can be interrelated within a large flexible system with the following properties:

- words can represent entities, actions, qualities, and relations
- words can be interposed with each other according to meaningful intrinsic syntactical relationships (syntax)
- words may share overlapping meanings
- words may build on each other in groups to form higher hierarchical levels of abstraction or new dimensions of meaning
- language may define terms and extend itself to provide comprehensive coverage of any scope of things including everything or nothing
- language has an unending potential for continued construction which may incur refined discernment, expansion, further hierarchical

abstraction, or new dimensions

• language may be constructed and used: for interpersonal communication, soothing, aggression, or play; or for personal or collective needs, planning, and adaptive integrations; or to model external reality as rational analytic thought, science, or philosophy; or to serve the higher sensibilities of aesthetics, value, humor, or justice.

Overall, Language provides reduced maps of anything and everything. It can capture and begin to understand and therefore deal with anything, and continually improve this understanding. The downside is that language inherently loses much of the directness, fullness, and richness of flavors of the original neurobiological representations, and must struggle to artificially represent those.

8.3.2 *The inner neurobiological roots of language*

At root, language is a structural devise to mediate between consciousness and the fuller ranges of neurobiological signals which underlie conscious apprehensions. The quadrangular relationship among consciousness, language, neurobiological signals, and meaning implied by this is indicated in Fig.8.1. The chapter has thus far dealt with the surface or upper level of intersection between language and consciousness which is the level of conscious meaning. Figure 8.1 now also takes us to the lower level interface of language with neurobiology, bringing us to its neurobiological bases. These neurobiological bases are of two kinds: those essentially non-linguistic neural represesentations relative to the element in question in itelf, referred to here as the 'referent', and those of language as applied to that referent (linguistic). There are inner neural representations (NB) as well as conscious experience (CN) of the referent in itself, such as visual and auditory images, all labeled here as 'core referent', and other neural representations and mentoexperiences of various associations of the referent, with, say, other features of the environment or internal valuations, judgements, or templates, non-linguistic and linguistic, all labeled here as 'referent-related'.

The linguistic representations are of two types: elements and larger organization structures called syntax. Both types include both conscious and neurobiological representations. The elements include letters, syllables, and words. These can be expressed or sensed and apprehended. They can be auditory, visual, or tactile, and used in auditory expression (including speech) and hearing (lingual), or writing and reading (literary). The organizational structure of language is generally labeled 'syntax', and has been referred to in the field of linguists as 'deep structure' in the last half-century. All of the linguistic and non-linguistic inner representations of language are indicated in Fig.8.1.

The syntactical deep structure of language is internal to language itself and has been described well by Chomsky[2] and subsequent linguist theorists. This syntactical ordering of language is what we consider a cognitive ordering. It does not necessarily have the same kind of relationships, dynamics, and implications as the original neurobiological ingredients which may involve sensory, motor, aesthetic, valuative, emotional, or existential qualities. Language has an inherent tendency to take us from the fuller sensibilities to the cognitive.

A word reduces a rich complex of distributed neuroelectrical and neurobiological activity across a wide collection of neurobiological neurons, regions, and interconnections to a single entity (the word) with a much simpler neurobiological representation. This word can represent a single entity or even a single simple encounter with that entity and implicitly intend to convey all the richness of the original entity or encounter. This gives us a fixed inner representation with easy relation to manipulative representation and higher abstracted representation. Moreover, other words can be constructed to represent any existential or secondary qualities associated with it. All this is very powerful and very useful. The price for this cognitive and communicative language is the continued reductive drift from the full sensibilities into the cognitive mode. The second manifestation is that even words which include or relate extracognitive qualities are secondary cognitive representations fully embedded in language which may help trigger but do not carry the rich flavors of the full original experience. Moreover, most words and higher order abstractions which refer to a central range of the neurobiology of the rudimentary images in question ignore additional

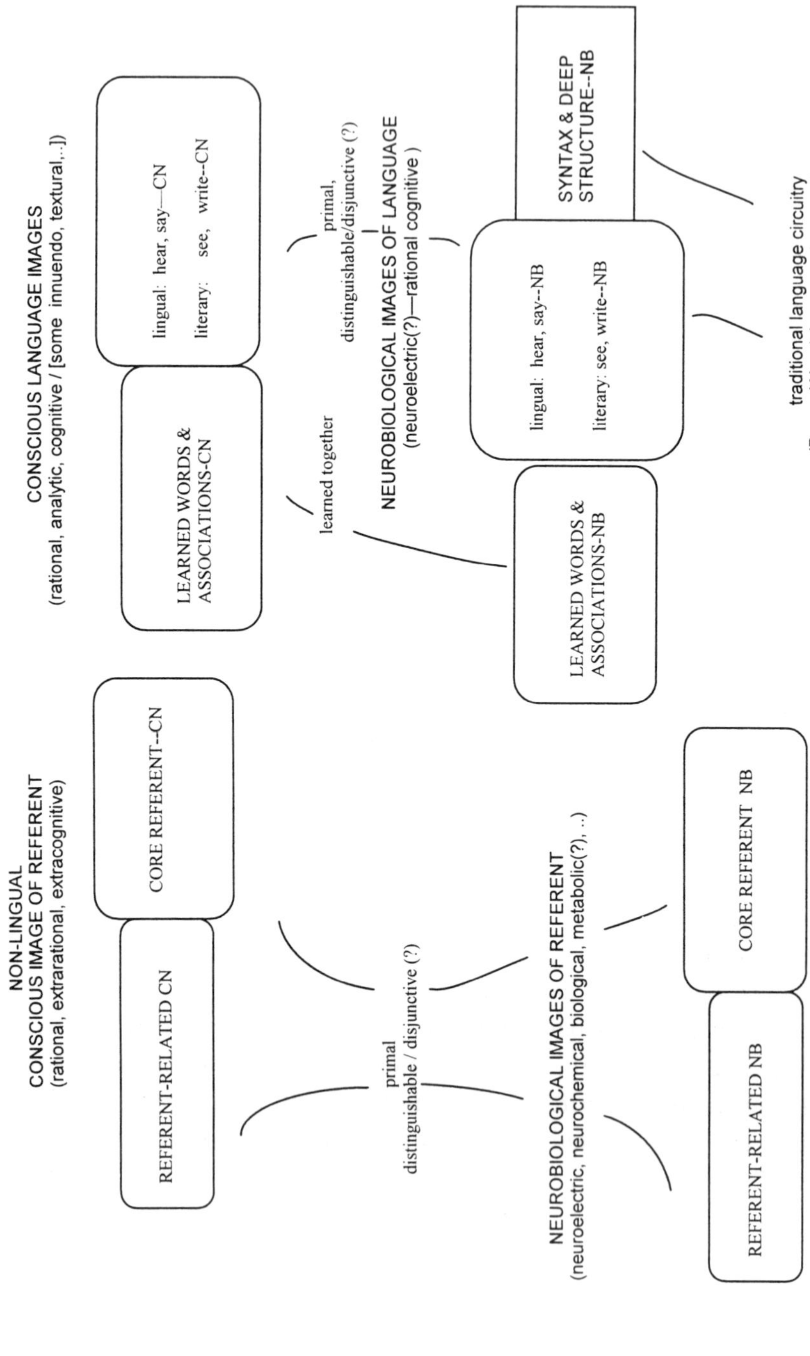

Fig. 8.1 Neurobiological ground structure of language.

peripheral resonances which are not necessarily part of the word's explicit meaning. Language gives us a powerful cognitive tool whose use leads us into cognition at the expense of fuller inner sensibility[3].

Nonetheless, resonances of these peripheral non-cognitive qualities associated with words, and the broader cognitive relations of words add innuendo and nuance to the use of language in any given instance. Relatedly, but separately, the resonances of especially the sounds of words and their relations in rhythm, rhyme, and assonance give spoken language an aesthetic quality which may lend elegance to speech and finds its fuller expressions in literature and poetry.

Altogether separately, peripheral cognitive associations tend to accumulate through sentences, paragraphs and sections to color any given discourse. This makes cross-cultural communications difficult.

8.4 The Conscious and Cognitive Use of Language

Consciousness uses language along with many other inner constructions especially in cognitive quests and communication. Most of consciousness, however, is not simply nor even primarily cognitive, and most of its inner constructions are best seen as neurobiological structures and patterns in themselves. Moreover, as indicated in chapter seven, one has the sense that even the quests of consciousness along the lines of what, who, which, why, and so on are in fact carried out by searches below the level of language itself, even though they may eventually manifest in language. The nature of cognition and its use of inner constructions is described in the next two chapters.

Endnotes

1. Penfield1975
2. Chomsky1972
3. Our bottom-up approach to the nature of language offers a viable substrate for top-down studies which indicate a larger neurobiological base such as:
   King et al1999 and Blanshard1939.

Chapter 9

First Principles of Human Awareness and Apprehension

This chapter considers the foundations of our experiential existence, awareness, apprehension, and sensibility; of their setting; and of how we know these and things generally. These provide the foundations for: the description of the inner sensibilities given in the previous chapters of this work; epistemology, the study and knowing of knowing and truth, which is continued in terms of neurobiological relationships in chapter ten; and our human views of our place in the whole of things which is taken up in Part C.

It is significant that we must initiate our discussions of the nature of human existence with statements of conscious awareness rather than the brain because only statements regarding the former (and only some) may be immediately recognizable and affirmable, whereas any regarding the latter are far from immediate human knowing--being obtained only secondarily and only with considerable collective effort. The idea that the brain may produce all of consciousness: is in itself an entity of consciousness; is judged only in such consciousness; and can be ascertained ultimately only by the self-grounding of all this directly felt conscious experience in the brain. For us, as humans, consciousness is primary, the brain is secondary: Experience precedes Brain.

This underscores the essential human centrality of our conscious existence, and the obvious fact that our neurobiology, as necessary as it in fact is, remains for most of us a secondary supportive structure for our meaningful realms of existence and living--indeed, one we for the most part lose sight of and can do without as long as things are going smoothly.

This chapter outlines what this work takes as a natural human process of pre-analytic apprehension which, when applied with analytical reason, can lead to reliable inherent understanding. When applied particularly to the essential features of our generic surroundings and condition, this leads to a foundational understanding of these in what we may label 'foundationo-rational' knowing. After this description, several important universally recognizable realms of foundationo-rational knowing will be introduced.

9.1 Pre-Analytic Apprehension and Inherent and Foundationo-Rational Knowing

The only possible foundation for our knowing is in direct conscious pre-analytic experience, for if we begin in any preconception we will be necessarily circular upon it. Pre-analytic apprehension provides a supportive backdrop for the more precise delineations of analytic knowing. In this, direct conscious experience is abetted by such processes as association, global arrangement, aesthetic perceptions, intuitions, and the like as described in chapter two. We will find some human foundationality here, and with a few universally recognizable premises, grounds for some fundamental human universality of perspective.

However, pre-analytic apprehensions are in the nature of ambiguous, uncertain, and often vague associations, innuendoes, maps, and estimates. Moreover, they are restricted to a narrow realm close to the tangible. When they stray too far into interpretation they become progressively less reliable, arbitrary, speculative, or idiosyncratic. To know more reliably, and more fully, over wider ranges, we find ourselves forced to rely on rational processes, either analytical or empirical, or both. Analytic reasoning builds upon premises. It is either ultimately grounded on postulates or essentially circular, somehow 'explaining' its own ground. Knowledge useful to human living must ground analytic reasoning within fields of pre-analytic apprehension. In this, rational analysis adds a circularly confirmable human objectivity, based on its intrinsic representational and evaluative rules, as well as a corresponding reliability which we call truth.

9.1.1 *Consensus, coherence, correspondence, and inherent and foundationo-rational truth*

Our foundational knowledge must be obtained from pre-analytic apprehension of qualities apparent in direct conscious experience. When such apprehensions are universally recognized in a consensus, they may be taken to suggest an 'inherent' truth. If such apprehensions particularly relate to essential features of our generic human condition and surroundings, and are recognized in consensus, they may be taken to suggest a 'foundationo-rational' inherent truth.

To determine whether such candidates for inherent truth, whether foundationo-rational or not, are indeed truths, they must further satisfy rational analytic criticism for consistency--coherence or correspondence, and, if scientific, also empirical tests for correspondence with experimental data.

Inherent truths may be suspected of representing intrinsic qualities of external reality as reflected in sensory representations or representing intrinsic qualities of our own neurobiology. If the latter, they may reflect intrinsic qualities of our rationality (as in mathematics) or intrinsic features of external reality as represented in our brains by templates such as archetypes. These latter may result from evolutionary modification of the genctic structure of our brain's networks as, for example, in instinctual perceptions or related fixed action patterns.

Pre-analytic apprehension, then, can provide a reliable foundation for our knowing of rudimentary generic features of human experience. We will now follow this approach to the three important realms of: human existence and apprehension, energization and human causality, and the contexts of things human.

9.2 First Principles of Human Existence and Apprehension

This section presents premises and associated 'first principles' grouped as intrinsic placement, the mind-body question, organic functionalism, foundations of human knowing, and a universal human perspective. All these areas, principles, and premises are graded from first through fifth 'order', according to their estimated strength of introspective universality. All have significant recognizability. These gradations gradually diminish in reliability from the direct experience of conscious awareness itself outwards, as it were, towards the margins of circularly-based rational analysis.

9.2.1 *Intrinsic placement*

We take as a foundational premise our unique individual functional/experiential existence as one with a partially conscious field of awareness. We take the presumed fact of our individual conscious existence as a first-order principle.

We take as additional premises the partial autonomy of our conscious awareness and its placement within the margins of broader unknown inner and outer realities, all inclusively referred to as the whole of things. These recognitions are based on further first principles of energization and differentiation. That is, we may be seen to differentiate our consciousness within a larger whole (differentiation) by virtue of the effects of our felt autonomy (energization).

We know of energization as the source of change which also implies time. Differentiation implies separable existence and therefore space.

This constellation of three first-order premises and first-order principles is the foundational area of intrinsic placement. We recognize this localization of consciousness within surrounding inner and outer realities in space and time as an intrinsic viewing point, and will refer to it as such.

9.2.2 *Mind-body question*

We take as a second-order premise the association of our existence in intrinsic placement with what we call our body, and this latter's mediation between our consciousness and the ambient whole of things. This is the second-order principle of mind-body association. This association is established by the constant contiguity of our awareness with our body, our persistent sensory experiences of it, and our autonomous expressive actions through our activations of it.

Another second-order premise is that of mind-body distinguishability. This reflects the recognition that both the content and existential texture of our conscious experience (associations, cognitive apprehensions, feelings, sensations) are distinguishable from the more tangible observable elements of our body and the external realms (earth, trees, flesh, brain).

A final second-order principle is inner sensibility. Its premise is that conscious experience forms part of a larger realm of inner sensibilities, which mediates between conscious experience and the body in some uncertain mix of experiential and physical qualities.

These premises and principles establish the foundational area of the mind-body question.

9.2.3 *Organic functionalism*

A third-order premise is that the inner sensibilities mediates organic wholeness among consciousness, the inner sensibilities itself, and the body. This is the principle of individual human organicity. It allows the recognition of one's consciousness, inner sensibilities, brain and body as an organically integrated self within the whole of things.

Another third-order premise is that the inner sensibilities mediates an overall operational functioning of the self within the ambient whole of things in some mix of: conscious awareness, feelings, wants; sensory perceptions derived from the body and surroundings; the guiding of inner and outer actions and expressions of the body; and an array of inner constructions, representation, evaluations and judgments. These are taken as a third-order principle of functionalism of the inner sensibilities.

The last third-order premise is that the inner sensibilities exhibits both physical and experiential manifestations, and that these two dimensions operate cooperatively or conjointly in much (or all) of its existential/operational functioning. This is the third-order principle of cooperative physico-mentoexperiential functioning of the inner sensibilities.

These three premises and principles comprise the broad foundational area of Organic Functionalism of the inner sensibilities.

9.2.4 *Universal human perspective*

We take as a fourth-order premise the recognition of other human beings distinct but kindred to ourselves within the whole of things. The corresponding fourth-order principle is human kindredness.

Another fourth-order premise is the inference of a constellation of consciousness and inner sensibility distinct but kindred to our own within every other human. This is a principle of separative multiplicity in intrinsic placement.

Implicit to this is the further recognition that we see the consciousness of others, not directly as we assume they do, but only infer it and relate to it from the outside and through the media of both their and our inner sensibilities. This recognition reveals an outer face of the principle of intrinsic placement, whereas its inner face is the first-order direct experience of our own unique conscious awareness described above. (These inner and outer faces may be distinguished by the terms 'inner sensibilities' and 'human sensibilities', respectively.)

A last fourth-order premise is the association of all human artifacts, both physical and conceptual--that is, all man-made things of any kind--and their differentiation together with all human beings as a distinct conglomerate labeled humanity within the whole of things. The corresponding principle is extensive human construction.

These fourth-order principles collectively produce what we can call a universal human perspective--the view of the whole of things from one's own intrinsic viewing point as consisting of: the inclusively nested self, humanity, and the whole of things. Clearly, such a view is universal only when particulars are recognized as idiosyncratic, and taken as variable contents of generic generalizations. This is taken up in Chapter 12.

9.2.5 *Foundations of human knowing*

These fifth-order premises refer to the general nature and circumstances of our apprehension which includes all our knowing. We take as a fifth-order premise that all our apprehensions, even those regarding the inner sensibilities itself, reside in elements which either: originate as part of the structure of the inner sensibilities itself, are constructed by it, or occur as responses in it to energizations from the surrounding whole of things. This is the fifth-order premise of physical-experiential apprehension.

Another fifth-order premise is that nearly all apprehensions of the inner sensibilities, even those regarding itself, are filtered, partial, incomplete, according to properties of the inner sensibilities, and, therefore, require confirmation, with the single exception of the existence of pristine conscious awareness in itself given by intrinsic placement. This premise implies a fifth-order principle of obscurity.

A third fifth-order principle is that large proportions of the activities of the inner sensibilities are outside of conscious experience (subconscious) including some highly significant cognitions and judgments, some in close operational relation to conscious experience and some not.

Another fifth-order premise is that humans inherently generate early in life by mostly pre-analytic apprehension a largely universally recognizable map of the large essential features of our generic condition and surroundings. We call this foundationo-rational knowing.

A last fifth-order premise is that some experiential human apprehension takes place according to existential/operational placement or functioning in terms of ambient surroundings. This implies a fifth-order principle of partial contextualization of human apprehension.

The first two premises together imply a fifth-order principle of foundational circularity in human knowing, that what we think we know depends on a vehicle which we can know only secondarily, obscurely, and incompletely.

The five premises collectively imply a broad fifth-order principle of epistemological pragmatism, inclusive of circularity, obscurity, partial contextualization, and need for additional confirmation.

9.2.6 *Comment on first principles*

This array of fundamental premises, principles, and question areas--conscious awareness, intrinsic placement, existence, energization, differentiation, change, time, space, inner sensibility, organic mind-body functioning, self, other people, humanity, surrounding external nature, sensation, autonomy, internal reflection--these, and others mentioned here or not, are the givens of human experience on earth. This work takes all these premises and principles as inherently accessible and clear to humans with a bare minimum of experience on earth--a universally common ground structure for grounding our existence and understanding, and for guiding the subsequent development of both.

Many readers will have recognized ideas common in western philosophy prominent in this statement of first principles--particularly those of Descartes, Kant, and James[1]. Yet, this work has selected from these and other thinkers only as stated here and within its own inner purposes, foundations, and direction. A different overall interpretation from any of these or other sources is implied here. Haack provides a view of the foundations of human knowing including a definition of a type of inherent knowing which we have adopted within our sense of pre-analytic apprehension[2]. Hacking and Redhead give discussions of the correspondence and coherence tests for truth, which we will discuss further in the next chapter[3].

9.2.7 *Pre-analytic knowing and language*

This chapter's considerations of pre-analytic apprehension suggest particularly that our apprehension is much greater than that of language, so that language seems both overrated and underrated in scholarly thought. For example, it seems certain that we and lower forms apprehend in a direct, non-lingual fashion, many basic concepts relating especially to our fundamental circumstances as creatures of the earth. This includes apprehension of the degree of possible threat to us by various objects, creatures, or others, perhaps depending mostly on how predictable their behavior seems. This is expressed succinctly in the poem in the frontice of this book as 'precarious things know about being precarious things'. On the other hand, language is underrated by post-modernists who criticize its capabilities for fine discrimination, whereas we can continually refine language with new words, and hence new concepts and shades of concepts to forge past existing barriers. This is considered more in Chapters 10 and 16.

9.3 Energization and the Inner Sensibilities

The principles of energization, or its equivalent, 'causation', and conscious awareness are among the deepest of first principles of human existence and experience. In this perspective this work sees our partial autonomy as a humanly vital quality of what must be recognized as a human causation with unknown relationships to other forms of energization, such as the ambient turbulences and forces of physical nature, plants, and animals; and the deeper ultimate forces in the depths of matter and origin of the physical universe and whole of things with all their obscurities.

The energetic quality of human causation encompasses and expands all of human motivation, from drive, homeostasis, pleasure, will, intentionality, and the like from basic biological to adaptive processes to objective and transcendent valuations such as truth, beauty, goodness, justice, wholeness, and excellence. The relation of our human causality to the physical forces which drive the neurobiological processes of our brains is fundamentally a question of energization. (See Chapters 4 through 7.)

9.4 The Contexts of Things Human

Pre-analytic apprehensions contribute significantly to the larger foundational outlines of our inherent contextual circumstances of existence considered in Part C of this book. The three basic contextual conditions of time, space, and differentiation are outlined here.

9.4.1 *Our existence in time*

This work takes the existence of consciousness and the intrinsic viewing point and the following qualities as universal to all individual human experience, given by the three premises of intrinsic placement: We exist individually locally within a space and time, both of which stretch out around us in multiple directions. We seem perennially placed in time in a moving present (becoming past) at the boundary between a past which seems irretrievable and a future (becoming present) which seems at least partially open and partially dependent on us. All our direct and immediate perceptions are realized here. We remain perennially at this boundary from birth to death. Our existential sense of this is our reality, no matter how our analytic cognition finds it best characterized, be it distinct linear and three-dimensional, or integral curved space-time.

In practice we live and operate mostly with internal reflections and representations of reality rather than with reality itself. In this, we exist in part as though distributed over time. We operate largely according to memories, circumstances, anticipations, and plans spread over spans of moments, hours, days, and weeks. This distributed existence is mostly internal and secondary, a step removed from both external reality and primary here-now existence. Children exist much in contact with the particulars of here-now experience. We exist more in terms of secondary internal templates (inner constructions) into which such particulars are absorbed often by-passing immediate here-now recognition. We come into more intense here-now experience in threatening or challenging experiences, adventure, play, or other remarkable encounters.

The inner sensibilities of valuative representation, discussed in Part A, bring us to internal realms of static representations where we can remove ourselves from time in transcendent involvements with the dimensions of aesthetics, reflective understandings, or value, or simply the space of being in and around its particular current contents.

It is interesting to point out that the brain's ability to integrate over time is rooted in the inertial time lags associated with its neurobiological processes. The most pervasive fundamental time lags in neuroelectric signaling are those required by the inertial rotations of the lipoprotein molecules of the membranes of all neurons, introduced in chapter five. These rotations produce the transmembrane neuroelectric potentials which are the brain's neuroelectric signals. These time lags are of the order of ten milliseconds. The neuroelectrical activity of the brain is fundamentally geared to integrate over time, and thus so are all its neuroelectrically-mediated apprehensions.

9.4.2 *Our place in space — the radiality of human existence*

The first principles given above have placed autonomous human organic functioning within larger realms of humanity, and nature. This implies a further fourth-order principle of radiality in our individual relationship with the surrounding whole of things. We are radial in the three senses of: (a) radially-centric--being intrinsically centered in one's particular brain and body, and the changing particular local circumstances of those; (b) radially-expansive--into one's surroundings by virtue of: constant sensory bombardment from external sources and their reverberations in our sensations and memory; of intrinsic bonds, tension, and coalescions with other people; and of practical utilitarian necessities of life imposed by our intrinsic biological needs and contingencies; and (c) radially-transcendent--according to the intrinsic internal transpersonal logic and judgments of the valuative representational qualities--perceptual aesthetics, cognitive apprehension and reason, and value and goodness.

In this tripartite radiality we are inherently involved with a multiple-leveled mix of concerns: personal, intimates, local and large-scale groups, universals, and transcendentals.

9.4.3 *Primordial differentiations*

Much of our internal cognition rests on internal maps of the whole of things with which we operate and place information. We will approach in Part C the larger contextual circumstances of human living in terms of a highly useful map of the whole of things as three overlapping realms of: the inner sensibilities (as of a given individual), humanity (other people and all human artifacts), and the whole of nature. The first of these comprises the inner contexts of things human, the last two our outer contexts.

This map is particularly useful because it: identifies the three distinct realms of energization which impact on our lives (all our personal energizations, those of other humans and collective humanity, and those of the whole of nature beyond humanity); readily maps onto our direct visual images of the ambient world we experience in daily life and our views of its inner workings; and provides a coherent basis for grouping all things and ideas. The central differentiations in this mapping are based on energization (in ourselves and externally) and kindredness to ourselves. This map is the topic of Chapter 12.

9.5 Further Remarks on Foundationo-Rational Knowing

This chapter has described a view of the whole of human knowing in terms of the two realms of pre-analytic foundationo-rational knowing and the rational methods of science and philosophy.

Apprehensions in the first realm arise within the pristine space of intrinsic placement, as pointed out famously by Descartes, but focused more narrowly than here. This work sees the realm as providing a universally recognizable cognitive view of our surroundings, our place in it, and our knowing about it and the view itself, as well as being the central fact of our existence. The approach produces a creditable vitally-important map of primary operative realms within and around us which are particularly significant to our existence and its energizations. This approach is existential and relies on direct personal experience and personal testimony.

The foundationo-rational approach soon runs out of clear universal low-order principles, and encounters a multiplicity of differing possible formulations. These gradually degrade in reliability from a pristine first principle of individual conscious existence to arbitrary differentiations or constructs. As the order increases it becomes progressively better to consider the constructs of the foundationo-rational approach as tentative pragmatic hypotheses within a rational analytic approach. In turn, these foundationo-rational principles themselves become subjected to the critical analytic inspections of rational analysis.

The first principles of apprehension described in this chapter are rooted in this foundationo-rational realm and provide the transition from it to rational apprehension. Rational apprehension is the way philosophy, discerning common sense, and when also empirical, of science. It is more adequate for the fuller understanding of external reality and its relationship to these foundations and to our lives in surrounding reality. Rational apprehension, as all apprehension, operates within our individual inner sensibilities and is dependent upon its structuro-operational processes. In this as well as in itself, rational apprehension is held by the first principles given above to be externally-oriented, foundationally circular, and dependent on external information and pragmatic for all but its inherent structural truisms (mathematics, for example). This is a dual circularity of chicken and egg, which can only be underscored, and perhaps somewhat clarified, but not transcended, by the interrelating of our knowing with neurobiology.

These kinds of considerations take us to the outer limits of the foundationo-rational approach to understanding. Chapter 10 looks more closely at the neurobiological aspects of apprehension and knowing.

Endnotes

1. Descartes1637, 1641; Kant1787, 1788, 1790; James1907
2. Haack1993
3. Hacking1983, Redhead1995

Chapter 10

Knowing and Neurobiology

This chapter relates human knowing to the neurobiological processes of the brain using its overall functional organization as outlined in the previous chapters. It characterizes knowing in terms of the patterns of inner constructions and perception. It establishes a neurobiologically-grounded typology of truths which provides new foundations for contemporary and traditional types of knowing. The approach opens up natural and more widely inclusive contexts for knowing and truths. It makes contact with important streams of contemporary epistemology and criticizes the foundations of post-modern thinking as is further pursued in the previous chapter.

This work sees human knowing as involved with: consciousness and inner reflective space, the pre-analytic foundationo-rational apprehensions discussed in the previous chapter, and with extensive plastic inner constructions especially in its higher analytic forms. Human knowing is characterized in this work by the first-principles of human awareness and apprehension, especially those fifth-order principles of human knowing itself and in these it is ultimately incomplete and pragmatic. The project of this chapter--the human conscious knowing of the neural foundations of knowing—is recognized here as inescapably circular but nonetheless highly instructive.

10.1 Inner Patterns and Knowing

This section will describe how knowing has to do with the relations of patterns within the inner constructions of the brain, and truth with match-mismatch comparisons between patterns. This view sees the evolutionary history of these inner constructions and their patterns against the watershed flowering in humans of plasticity, consciousness, imagination, and self-direction (Chapter 7) which brings on a major departure from the earlier dominance of genetic prescription for much of behavioral living. Genetic prescription underlies the structured forms of recognition by archetypes and instinct with its trigger stimuli and fixed action patterns (shared, for example, with fish, reptiles, and birds). The coupling of these with a moderate degree of plastic learning gives rise to an appetitive behavior characterized by the recognition of incentive stimuli and action-contingent situations. These may characterize most of mammalian behavior. The vast increase and freer higher order of plasticity in humans, as the defining characteristic of our species, exceeds these more limited manifestations in lower vertebrates. It manifests in humans as an extreme increase of internal representations, imagination, and active plastic inner constructions, and remarkable increase in external constructions of tools and adaptive behavior. Many different self-consistent subsets of inner constructions have been indicated in chapter eight; they include language, mathematics, science, history, stories, and anticipations, scripts, plans, and intentions. Many, but by no means all can be considered adaptive integrations.

Rational analytic knowing is seen here to operate independently but normally within circumstances indicated by pre-analytic apprehensions such as associations, arrangements, and estimates, what this work has collectively referred to as a generalized cognitive capacity. This capacity normally operates within the larger reaches of the full inner sensibilities including generalized aesthetic perception, especially with its intuitions, and value, adaptive integrations, and biological primals. This cognitive capacity, and particularly its rational analytic mode, is the source and home of truth. Rational cognitive patterns are focusable and can be freed of extraneous factors and self-concerns. In this they can provide objective truth. Conscious partial autonomy guides attention internally into the banks of inner construction to prompt integrations, comparisons, and the further development of these and the creation of new constructions. This is what we call thinking or reflecting (see

Chapter 7). Such reflective thinking comprises a quest which results in the extensions of knowing, the creation of additional understanding. This understanding becomes an integral contributory part of the underlying fabric of one's existence and behavioral living. This reflective thinking is a central manifestation of the watershed plasticity of our species.

Knowing and the relative truth of knowing are central and pervasive features of human life. Most of us know a little about: other people; the entities and processes of nature; a few arts, sciences, and very, very many adaptive skills, small and large; the institutions and ways of our communities and nations; our dangers, wants, preferences, and opportunities; our neighbors and our way around. Collectively, we have major academic fields which embody our most rigorous quests to know: first, mathematics and logic, then, science and philosophy; then others which seek wider, if less certain knowing: literature, the arts, religion; and a myriad of professions and vocations with ever-increasing useful specialized knowledge across unbounded human interests and needs. For the less austere forms of knowing, both knowing and its truth can be partial and graded. For the more austere--mathematics, logic, philosophy, science--more severe tests must be applied to winnow the purely true from the questionable or extraneous.

We know a thing if we have an internal pattern of it, an internal pattern which is almost but not quite always fixed in the sense of being recallable in the same form. The pattern itself may be static or include temporal change. Our knowing is true if this internal pattern matches the thing. Things that can be known are things that can be themselves represented by patterns in the nervous system--patterns which in this case may be transient sensory perceptions or qualities or associations of these, or inner constructions of any kind, including, for example, derivative impressions, images or thoughts, or more highly structured composite patterns, as in language. Things and their patterns may be simple, unitary elements but more often are multiple groups of patterns in some kind of arrangement--for example, a middle-sized theory in science, or the best way to see a difficulty in one's life.

Estimates of truth thus may entail two levels of matching tests: one for the coherence within the fixed inner patterns themselves, another for the correspondence of the fixed inner pattern to transient patterns of the things of interest. Tests for coherence estimate the internal integrity of a pattern. The most rigorous form of coherence demands the complete absence of contradiction and establishes a mode of knowing and truth which we call analytical or rational. Traditional western philosophy is a good example of analytic rational truth. Tests for correspondence examine the relationship between two sets of patterns, say inner thinking on the one hand and externally-derived information on the other. This is best exemplified in the experimental testing procedures required in science. We call this correspondence testing empirical knowing, and the approach which uses both coherence and correspondence testing analytic empirical knowing. Haack has given a highly useful view of the foundations of human knowing and a definition of a type of inherent knowing which are highly akin to our sense of pre-analytic apprehension[1]. Hacking and Redhead provide penetrating discussions of the correspondence and coherence tests for truth[2] especially within science.

This interpretative picture can lead us to specific neurobiological ground of knowing and truth in the brain. The patterns of inner construction may relate to any capacities of the inner sensibilities and take a wide variety of neurobiological forms. Let us suppose as a working hypothesis that the patterns of analytic and empirical knowing in particular relate to multiunit neuroelectric patterns in primarily neocortical neuronal networks or coalitions of networks, and that the sculpting and shaping of these patterns is effected by the underlying complex multiunit patterns of synaptic interconnections. Suppose further that the processes of coherence and correspondence accord to appropriate match-mismatch of subsets of these neuroelectric patterns within (coherence) or between (correspondence) groups. Contradiction, for example, would be flagged by a mismatch of patterns. Such a model give us not only a highly suggestive picture of the brain's involvement in knowing and truth, but also the basis required to make quantitative dynamic modeling of the hypothesis. This procedure is a powerful working tool, necessary for the deeper-reaching sharper understanding of this complex field. The approach lends itself to the exploration of higher-order governing biological principles such as the concept of cognitive dissonance as a major operational and motivational factor in human understanding[3]. Our view would see cognitive dissonance to manifest as mismatches in the coherence and correspondence tests between multiunit neuroelectric patterns. Such patterns can be and have been studied with neuroelectric experimentation and associated neural network models and computer simulations[4].

10.2 A Typology of Knowing

Table 10.1 sketches a typology of knowing according to whether the underlying neurobiology is genetically prescribed (hard-wired) or plastic (learned or constructed), and whether the knowing in question is (nearly) universal or universally replicable across people, or is limited to individuals or a particular group. This approach reveals three types of intrinsic knowing, all of which are genetically prescribed (hard-wired), prior to experience: cognitive intrinsic (knowledge true by internal identity--tautologies such as mathematics and logic; for example, 2 + 2 = 4 because both are different names for the same thing); archetypical knowledge (neural templates or associative connections relating mostly to instinctive perceptions and behavior; for example, recognition of mating partners or species enemies), and idiosyncratic intrinsic (genetic mutations or singular abnormalities in individuals or small familial groups; for example, genius in itself or mentally-challenged computational genius).

Cognitive and idiosyncratic intrinsic knowledge are both what Kant would call analytic a priori knowledge. Our archetypical knowledge is an interesting extension of traditional epistemological classification as a form of synthetic a priori knowledge wherein genetically-prescribed brain patterns effect internal synthetic associations relative to perceptual encounters. These would be the encountering and plastic engagement in personal experience of intrinsic internal templates of archetypes as in instinctual recognitions described for example by Jung or the ethologists[5]. Kant saw synthetic a priori as a means whereby philosophical knowledge could transcend experience. For example, the statement ‘every event has a cause’ could be seen to imply a first cause[6].

Such synthetic a priori thinking is not widely held today. It is quite different from the phenomenon suggested here.

Table 10.1 shows four types of plastic knowledge based on whether or not they reflect winnowing by the testing processes of analytic coherence or empirical correspondence or both. Universally recognized plastic knowledge may be suspected of reflecting intrinsic qualities of external reality. The most widely recognized example of this is the analytic and empirical knowledge of science. This is especially our most solid practical knowledge, pervasive particularly in western culture within the last century or two. Another highly important example, less widely known, is the universal pre-analytic knowing we call 'inherent' after Haack[1]. A salient example is the human foundationo-rational knowing of essential features of our generic surroundings and condition introduced in the previous chapter. This kind of knowledge is essential to human existence and survival yet is so common as to seem slighted in broad philosophical assessments of our understandings of things.

Non-empirical analytic knowledge is rationally analytic and coherent but need not reflect sharp correspondence with reality. Different individuals or cultures may differ widely in their regard for any given specimen of it. Many particular philosophies or theological views are of this type.

Non-analytic plastic knowledge, which is neither analytically nor empirically grounded includes by far the greater part of human knowing. It is the mainstay of our individual and collective living, containing most of our conceptions and interpretations of the human worlds in which we live. These inner constructions (indicated in Chapter 8) are the defining characteristic of our species, and generally highly robust. Nonetheless these are, unless carefully weaned, typically riddled and interwoven with many uncertainties, overlooked contradictions, shortnesses of scope, personal preferences, local biases and the like so that, useful as they may be, their universality may be wanting and their truth levels unreliable. This content may range from sagacious wisdom, highly robust individual or collective adaptive integrations, to gratuitous conjecture, spurious half-truths, or short-sighted provincial bigotry.

It is interesting to note that much of the genetically undergirded knowledge we have labeled 'intrinsic' may in fact be seen as higher-

Table 10.1. A typology of knowing.

	INTRINSIC (genetically-prescribed)	PLASTIC (learned or constructed)
UNIVERSAL and UNIVERSALLY REPLICABLE	COGNITIVE INTRINSIC (mathematics, logic ...) ARCHETYPICAL INTRINSIC (neural templates, instincts, ..)	ANALYTIC EMPIRICAL (science, ..) INHERENT (corroborated foundationo-rational ..)
VARIABLY CONTINGENT or LOCAL	IDIOSYNCRATIC INTRINSIC (genius, psychopath, ..)	ANALYTIC NON-EMPIRICAL (philosophy ...) NON-ANALYTIC PLASTIC (arts, literature, folklore, conjecture ...

order inherent species knowledge of sorts in that it is the result of selected genetic alterations to earthly environments.

10.3 Knowing and Contemporary Thought

The first paragraphs of chapter nine argue for the primacy of direct pre-analytic apprehension as the natural supportive contextual background of most rational analytic thought--pre-analytic apprehension being taken to include association, global arrangement, various obscure archetypical templates, aesthetic perceptions, intuition, and the like. Universal pre-analytic apprehension contributes centrally to the concept of inherent knowing recently described by Haack (1993). This chapter defines inherentism as a realm of knowing which depends on a combination of pre-analytic and analytic processes and consensus, and suggests truths which may reflect intrinsic qualities of either external reality or of ou own neurobiology. The recognition of the universal place of inherent knowledge in human life loosens the tight epistemological squeeze of hard rational analysis in itself, and therefore weakens the denigration of human knowing advocated by post-modernists. In this view, rational and empirical analytic thought, the focus of most epistemology, operate normally within this bedwork of pre-analytic thought which in turn operates within the larger whole of one's inner sensibilities, which is operative within larger surroundings--natural, intimate, and sociocultural. We have projected further that rational analytic thought may indeed operate independently of all these surroundings in passive reflection and according to its own intrinsic processes of valuative representation. In this rational analytic thought defines truth for us individually, and serves as a realm of objectivity and a mode of transcendence for those who will hear and adhere to its full nature.

Generally, plastic contingent knowledge has been held in low regard in philosophy at least since Kant. This contributes significantly to the nihilism, relativism, and general pessimism of much post-modern philosophy of the the last half-century or so. The pragmatism of William James bespeaks a much stronger recognition of this kind of learning in human affairs. Indeed, this work sees this pervasive higher plasticity as the defining characteristic of our species. Plasticity is our game plan for survival. It, with its companions, imagination and creativity, are the reason we are, by huge distances, the dominant form of life on the planet. This plastic knowledge is strong, robust, comprehensive, adaptable, flexible, creative, prolific, and self-corrective. The ideas of dependence on contextualization and conditioning and personal radiality of perspective seem here as highly exaggerated in post-modern thought; this work holds that our cognitive capacities intrinsically operate according to objective criteria and the patterns on which they operate can be freed from self-interest easily enough by any responsible person. Similarly artists are not slaves to their early conditioning nor material circumstances, as many post-modernists would have it. To the contrary, many creative people find their liberation precisely within the transcendent objective valuative worlds of our higher sensibilities of aesthetics, knowing, and value. In addition to the objectivity available to individuals, candidates for collective policies or truths can be, and are, subjected to tests of universal replicability or recognition. Self-interest does sometimes muddle things, but sometimes it should play a part, and often it is not involved or is overcome by the larger transpersonal concerns which all individuals also possess to some degree. Humanity is fundamentally in this mix of things (as discussed more fully in Part C). It is troublesome, but not fatal--certainly no more than life itself. One should note that the adaptively integrative plastic pragmatic knowledge of individuals, including but not restricted to valenced thoughts, idiosyncratic as it is, is exceedingly, indeed preeminently, important to personal, local, and regional living and interests. Moreover, such personal cognitive integrations point to underlying undercurrents of the human condition which are the ground stuff of wisdom, psychology, and literature.

This work holds that language is often overrated in epistemology. Chapter 8 has described language as an intermediary between consciousness and fuller inner neurobiological representations, cognitive and otherwise. In this, for example, the text is not all there is (another favorite cliche of post-modern literary criticism). There is certainly a realm of fuller human sensibility underlying significant literary texts of which the language is an intended indicator. The reader reads to find access to this underlying sensibility in himself to the degree that he can. That this may not always be completely effective in no way diminishes it as the essence of the project. Repeated reading increase one's sense of this depth. Further, at any stage of understanding, language itself is only approximate, but can be ever refined by thought and the creation of new words. Language may be sensed according to valuations of truth or aesthetics. Ultimately language is only one vehicle of sensibility geared mostly to cognition and communication. Other forms of sensibility play important parts in each of these.

This chapter has rounded out an inescapably circular task --the grounding of a rational comprehensive view of the foundations of the rational cognitive properties of the inner sensibilities within the inner sensibilities itself. Despite this feature the chapter has nonetheless outlined a comprehensive theory of knowing which is subjectable to both rational criticisms of correspondence and coherence, communal consensus, and the empirical testing methods of science. It has outlined relations of the mechanics of this field to operative neurobiological processes of the brain, thereby bringing epistemology closer to potential contact with the empirical testing procedures of physical science.

Any cognitive view of the origin or nature of knowing, such as epistemology, post-modern thinking, or any philosophy of mind are all fundamentally in a bit of chicken-and-egg circularity. A theory that relates cognition to explicit brain processes may be ultimately circular as well, but does point to a wider universe of potential ultimate grounding (partial or total) in physical nature.

Endnotes

1. Haack1993
2. Hacking1983, Redhead1995
3. Festinger1959
4. Gerstein et al1978, Abeles1982, MacGregor1993, Poznanski2002
5. Jung1953-61, Ethology
6. Kant1787

Chapter 11

The Larger Nature and Setting of Value

This work takes a wide scope for value which includes the good, the wise, the religious, and the moral, but also many other dimensions as well, such as loving interrelationships, fellow feeling and empathy, senses of personal meaning and accomplishment, basic creature comforts, contentment, bodily and spiritual vitality, fun, humor, play, indeed, all values. This chapter makes further interpretative comments on the foundations and wider nature of values as introduced in chapter three, and outlines the foundational relationship of these to the fields of philosophical ethics, wisdom, and religion. Philosophical ethics is seen here as a construction of reason within the cognitive-reason capacity itself, and both wisdom and religion as realms more directly associated with value and with dual-process valenced thoughts beyond cognition in itself. Chapter 14 returns to this subject following intervening material on man's external circumstances and condition. The tables of this chapter are important to this work because they identify central human values recognized by philosophy and religion thereby rounding out the description begun in Table 3.1.

11.1 The Essential Nature and Structure of Value

Human value is an intrinsic capacity of the human inner sensibilities. Human value is projected into surrounding reality and other humans by our inherent radiality and our actions.

Value is the foundational sense of significance and meaning in things human. It guides, motivates, and enriches experience.

Value is an independent dimension in itself. Ultimately all values must be judged according to the dimension of value within the sensibilities of individual humans.

Although many values are intrinsic and inherent, value is inherently at least partially self-corrective, in that all value may be interwoven and balanced out with valenced thoughts which inherently recycle and self-modulate.

This view of value and its structures is value-neutral in itself. Particular values and all weightings of particular values and the balancings between them, which are the content of these structures, are not.

Human value is an intrinsic independent operational capacity of the sensibilities of each individual human being, distinct but cooperative with the other capacities of reason, aesthetics, personal adaptive integration, and biological primals. It is deeply associated with the overall well-being of one's self and of the whole of things, inclusive of all one's sensibilities, external interrelations, circumstances, and contexts, to all that tend to enhance or threaten these, and to all apprehensions, reflections, and anticipations. Elements of value are intrinsically grounded in the fundamental multiple dimensions of the inner sensibilities--physical, physiological, neurobiological, and mentoexperiential, as embedded within total reality, and often somewhat obscure.

Three main classes of value are so rooted in the operational structures of the human inner sensibilities--intrinsic and sociopersonal values, and valenced thoughts. Intrinsic values are directly reflective of multidimensional sensibility processes including their neurobiological roots. They include: the essences of valuative representations, intrinsic personal values, and intrinsic interpersonal values. Sociopersonal (mixed) values are intrinsic personal and interpersonal values in learned plastic linkages with particulars of one's personal experience. They are particularly germane to rudimentary interpersonal and social cohesion. They are the locus of much tenacious early conditioning. Valenced thoughts are cognitive representations which are charged by association with the ground structures of intrinsic values. They include: socio-universal and general abstract values, and wisdom.

The essence of all human values (intrinsic and mixed values, and valenced thoughts) is a separate, independent dimension within the sensibilities; this essence is outside of and beyond rationality and knowing. Values, including valenced thoughts, can only be suggested, but never derived nor ultimately judged within rationality (except only with respect to their rationality), neither from first principles, nor anything else there. This observation provides resolution of the Faustian malaise, which is a despair that value is not realizable in reason alone. The 'logic' of value is not rational nor sequential, but rather direct, immediate, and associative--even circular or helical.

The natural context of human value resides in the sense of significance it gives to human experience, and the corresponding guidance it offers for human choice and self-direction. We may translate this into a partial guidance of adaptive integration and biological primals by valuative representation particularly inclusive of value. Value judgments are made in the confluence and balance of a wide collection of variably weighted individual value elements, not peremptorily programmed to most particulars, but sensitive to the overall balance of the whole, including all contingent circumstances. Judgments may be made ever new--fresh, unique, particular, often idiosyncratic, and often set in unknown or partially-known circumstances, according to various and mixed purposes and motivations--although people all too often uncritically apply old conclusions to new situations.

Value has a strong intrinsic relation to personal, purposeful, behavioral, and overall integration, as well as to biological primals. This milieu, the global, all-inclusive nature of value, and the intrinsic internally-pressurized triad of radiality (centricity-expansivity-transcendentality) encourage a range of human concern and energization across at least five distinct overall operative levels: personal; intimate; local groups or collectives; universals; transcendentals (as aesthetics, reason, goodness) . To the extent that concerns with the personal, intimates, or local groups predominate, particularity, diversity, and often cross-purpose and conflict will rule. To the extent that universals and transcendentals engage across sensibilities, common universality can be sought.

Any human instance of value judgment typically involves the balancing of many and often conflicting participating values. A typical instance engages a large heterogeneous grouping of intrinsic, plastic, and mixed types of individual values, all of which must be interbalanced within and across the circumstances and individuals involved in the instance. Natural contextual domains may involve individuals, intimate or local groups, communities, and cultures, and relations within and across these. Common intrinsic values are often more easily resolved than the middle and higher levels of plastically-influenced values. Degrees of ignorance, learning, and understanding or their absence, and of training and conditioning, for good or ill, both play highly significant, often leading roles. Individual codes of living; local, regional, or cultural mores; codes of law, religion, wisdom, morality, all reflect and may contribute significantly to these.

11.1.1 *Intrinsic values*

Intrinsic values are the ground structure of human existence. They include a wide range of immediately recognizable qualities. Most people would regard at least some subset of the following as intrinsic values: interpersonal attraction or aversion, trust, love, kindredness, empathy, compassion, envy, curiosity, animal vitality, fun, humor, appetites, bodily pleasure, satisfaction, contentment, happiness, fulfillment, melancholy, pain, want, sorrow, grief, anxiety, fear, anger, peak experiences, truth, beauty, goodness itself, achievement, justice.

11.1.2 *Sociopersonal values*

Sociopersonal values include many important rewarding and useful bonds, tensions, and coalescions relative to one's personal intimate and social circumstances. Sociopersonal values are formed by the association of particular instances of the common interepersonal experiences of human living with intrinsic values, willy-nilly from the daily experiences and hearsay of one's life. Such particulars are often tenacious. They often include various world-views shared within local groups or cultures. These are often of limited scope, and become sources of volatility and conflict. Resolution of these depends on distinguishing the plastic from the intrinsic, and integrating the particulars with the higher socio-universal values in the processes of Popperian balancing and growth (see below). Such cultural growth is difficult and slow. Constructive growth of sociopersonal values in individuals can be facilitated by increased life experiences in broader and richer environments.

11.1.3 *Valenced thoughts*

Much content of value, particularly at the outer ranges of human activities and frontiers, is in the realm of valenced thoughts, beyond genetic pre-programming and intrinsic value--open-ended, and often within our individual and collective purview. Valenced thoughts are plastic pragmatic values--tentative, contingent, malleable, open-ended, pragmatic half-truths, which must operate in confluence and balance with other values to find their own levels within particular partially-known circumstances.

Valenced thoughts have the same tentative and contingent nature as do theories in science. They are valid only on the whole and up to the present time. They need to find validation, correction, evolution according to how they play in experience. They must be validated according to their roots in both reason and particularly value. As with scientific theories, it doesn't matter where valenced thoughts originate, it only matters 'how they play'. They do not need to be derived within reason or past experience. They only need to pass tests of validity and range. Validation can take place only in the ultimate seats of judgment within individual humans.

In Popper's powerful theory of scientific progress[1], useful but incomplete principles actively prompt their own supercession, and are superseded or absorbed into larger more comprehensive principles or systems, often retaining their limited reaches of usefulness within the new view. The most useful theory is often one which most directly points to the seeds of its own supercession, corresponding to advancement in science. Popper's theory also can be seen as applicable to valenced thoughts as well. Popperian growth may be seen in the continual self-correction of many plastic pragmatic systems of human life--by refinement and expansion, or by replacements such as disjunctive paradigm shifts--according to ongoing reconsideration of judgment in individuals and collectives. This view stands as a constructive interpretative view of the inherent growth of human understanding which contrasts sharply with the reductive and pessimistic post-modern view which stresses the vulnerability of individuals to their self-interest and sociocultural conditioning in both science and life.

Principles and complexes of valenced thoughts, like all things human, are mixed, partial, limited, and dependent on individual peculiarities of experience and training. To guard against maladaptive learning and conditioning, the fundamental plastic processes of the inner sensibilities may be seen to continually recycle, and thereby refilter and self-correct all products of conditioning and learning through processes of value and of reason. Thus, both value and reason are resistant to inadequate learning and conditioning, even though early and highly impressed instances can be extremely tenacious. Any given human set of mind or interpretative perspective is also subject to these intrinsic processes of plastic self-corrective restructuring. Over-enclosure and provincialism can be corrected only by richer, more extended experience or powerful internal restructuring.

Human value is ultimately self-sufficient, with ever-refreshable individual revalidation and possible further refinement at hand for any particular value, no matter how it was discovered--just as gravity or electromagnetism have been discovered, then increasingly understood.

In all the workings of valenced thought, as in any processes involving our intrinsic plastic cognitive capacities, three distinct factors may be involved: genetic ('nature'), environmental experience ('nurture'), and partially free autonomous individual construction ('construction'). This last is fundamentally rooted in the fabric of our watershed plasticity (Chapter 4).

11.2 Ethics and Moral Philosophy

The rational foundations, nature, and qualities of human value and morality are considered in the philosophical field of ethics (also known as moral philosophy). Table 11.1(a) shows a representative array of values considered throughout the long history of this field[2]. This table is particularly important for this book in its illustration of human values recognized in academic philosophy. It rounds out the description initiated in Table 3.1. Table 11.1 a juxtaposes various aspects of values (particular values, ethical judgments, scholarly reflections) to produce an impressionistic image of the scope and nature of this field. (These disparities across entries help suggest the larger scope of material beyond the table itself. There is no pretense here of engaging this subject in a rigorous fashion in itself.) The ethical values of Table 11.1 (a) are grouped according to their strongest relations to the value groupings of: valenced thoughts, sociopersonal values, and intrinsic essences of the valuative representations. Philosophical ethics identifies essential origins of values in individual and interpersonal well-being-- affectional, rational, and behavioral-- as can be seen in Table 11.1(a). It is general less explicit, however, regarding attitudes of the heart and religious values.

Philosophical ethics constructively advises us as to the foundations, perspectives, contexts, coherence, nature, and comparative qualities of values, virtues, and ethical views, as seen in rationality. For example, in the last century it has illuminated fundamental rational weaknesses in the dogmas and postures of western religions. [See last subsection in Table 11.1(a)]. Although temporarily dissettling to some, these insights encourage a progression to a fuller, more accurate confluence of our reason and mentoexperiential capacities, and a better understanding of our common humanity across individuals, beliefs, and cultures.

Although ethics and rationality identify extra-rational origins and qualities of values, they do not provide for the generation of value in itself except in terms of purely rational characteristics and rational values themselves (truth, coherence, and the like). This can be labeled the 'Faustian Conundrum': one cannot get to value from reason alone.

It is significant that fuller extrarational values and attitudes themselves are often the primary concern of the deeply religious, as is indicated later in this chapter, and that these are not substantially included in philosophical ethics.

11.2.1 *The philosophies of religion and value*

The field of the philosophy of religion deals primarily with the overall settings, contexts, and purposes of religion, and the rational meaning and justification of religious claims (mythology or dogma) and values, rather than with their values as such. The The philosophy of value considers questions about value, such as: the meaning of being a value, whether value is an objective or subjective attribute, and what things have value.

We see value as a primal dimension, ultimately impossible to define except in its own terms. We see value as an intrinsic capacity of the individual inner sensibilities, subjectively recognizable within ourselves, and objectively inferable in others (as the force of gravity is inferable in nature). What things may have value are legion; they do so by virtue of their relationship to the judgments of human sensibility; the types of value are several--intrinsic (personal, interpersonal, valuative-representative-essences), value-associated sensibility elements of any kind (for example, sociopersonal); and value-weighted ideas (valenced thoughts).

11.3 Wisdom

Wisdom is a highly important realm of human understanding, which is generally not systematically cultivated. Wisdom is seen here as a dual-process product of the rational and value capacities of the inner sensibilities. It has usually been included peripherally, if at all, in western philosophy as such because of its implicit interinvolvement with values and circumstantial contextual settings.

Table 11.1(a). Representative values of philosophical ethics — Part 1.

broad plastic pragmatic virtues--metaethical or metaphysical
-follow the human purposes natural to our earthly circumstances
-harmony of action among the parts of the soul--reason, appetites, and spirit.
-one's rational 'good will' in adherence to duty for its own sake, beyond self-intersest --benevolence
-act so that one's maxims be universal laws ('categorical imperative')
-the greatest happiness for the greatest numbers of people
-qualities useful or agreeable to ourself or to others
-the good is not natural but is nonetheless metaphysically real and intuitively knowable

sociopersonal values
-temperance, courage, justice, and prudence
-pleasures of human intercourse
-keeping promises
-integrity, honor, honesty, loyalty, moral courage
-inherent interpersonal war--governments are essential to protect us from each other, rights

intrinsic personal primals
-self-interest is intrinsic to humans
-human passion as rooted in desire for self-good
-rooted in human pleasure over pain
-happiness and flourishing
-pleasure and evolutionary fitness

intrinsic interpersonal primals
-self-love or benevolence
-affections and benevolence are also intrinsic goods
-'fellow-feeling' (compassion, empathy)
-all persons should be treated as ends in themselves
-avoiding abuse of others
-religious virtues

Table 11.1(a). Representative values of philosophical ethics — Part 2.

valuative representation-essences:(aesthetic, value, reason in themselves)

the aesthetic in ethics

-enjoyment of beautiful objects

value in itself

-deepest attitudes of heart and senses of ultimate meaning

reason in ethics

-reason or sentiment
-reason acquaints us with moral duties
-acting in accordance with reason
-feelings and emotions are guided by reason so that behavior is appropriate to situations at hand
-reason is necessary to instruct us as to the results of our actions.
-rational analysis of meta-ethical questions such as semantics, objectivity and verifiability, and skepticism

critical metaethics

-universality attacked--wide diversity of values across world cultures
-foundlity attacked--epistemological difficulties, weak in extra-rational, context-dependency
-likes and preferences, merely describe emotions, merely introduce an imperative
-intrinsically plural
-inner motives and states of individuals
-concerns with one's ground projects and commitments
-ways of acting to make up a good life
-fact-value distinctions and 'is' and 'ought'
-free will as opposed to determinism
-'how much morality is it reasonable to expect of humans?

This work sees wisdom as a domain of valenced thought characterized by a comprehensive inclusiveness in the knowing about (that is, understanding) the living of human life. Wisdom is a learned plastic constellation of cognitive elements with explicit or implicit value weighting. In an austere abstract form, the truth level and scope of the elements may be the ultimate value. Usually, though, the underlying value is the larger concern. Wisdom, value, and consciousness share an inherent kindredness in their mutual association with the living of life and with the inner constructions which lead it. In this, wisdom may be seen as the deepest value-weighted knowings of consciousness.

In content, wisdom seeks understanding of man's fundamental nature and place in ambient reality, and the founding of conceptions of the good life reasonably confluent with that nature, and even perhaps the partial utilization of some of them.

Wisdom may play a variable place in one's life, from a variable means of self-protection, to all-encompassing commitment.

Wisdom systems are always partial and incomplete. This follows from our smallness and limitations before total reality and its obscurity, and from the nature of wisdom as a complex of inherently limited valenced thoughts as discussed in Chapters 9 and 10.

Any wisdom system that considers itself whole is moribund and must be transformed to a plastic growing system to retain validity. In wisdom as in science it may often be that the most fruitful truths are those that lead most directly to their own subsequent corrections or incorporations within fuller, more inclusive and broadly reaching truths.

There are many valuable pieces of wisdom within the religious, aesthetic, literary, philosophic, and common traditions of the world, and the human sciences (some might say overwhelmingly many). Much is so imbued with particularities, uncertainties, or biases of presupposition, context, or range as to be difficult for individuals to either place, access, or apply. This work sees the particular pieces of wisdom offered in all these or any sources as candidate elements to be subject to the continually recycling processes of Popperian validation, balancing, and self-corrective growth within the fuller personal thought systems and ultimate seats of judgment of individual humans. In this, our view offers a pathway for the multitudinous growth and synthesis of the varied wisdoms of the world in the ongoing experiences, reflections, and intercommunications of its individuals.

Perhaps the deepest wisdom resides in accurate senses of: the nature of the undercurrents of human existence on earth and in social interrelation; one's deepest intrinsic nature and place with regard to these larger surrounding currents; and judicious choices as to alignment-engagement-disengagement with those currents.

The particulars of wisdom systems are conditional, contingent, context-dependent, and elusive. A world view in any environment rests on a fabric of many unspoken, often unrecognized assumptions, upon which the world view is a collection of derivative projections. It is difficult to know whether a world view is successful in its own merit or simply because its holders are accepted or successful in a particular culture or environment, perhaps for totally different reasons than their supposed wisdom. If one's environment changes, the relied-upon world view may no longer correspond with well-being or an adequate interpretation of surrounding events.

The highest human wisdom may reside in knowing how to behave in situations one doesn't understand. Inner confidence and depth of faith are often determinative factors in the durability of one's world view.

Wisdom often consists more in how to manage or relate to things than in the internals of things. It often seems more compatible with useful senses of the overall relations and contexts of things rather than with analytical understanding.

Wisdom abets human living; a rudimentary amount is necessary for worldly survival. Yet, wisdom in itself does not insure happiness, success, recognition, or acceptance, and does not provide but may help guide one to ultimate senses of meaning, purpose, and fulfillment.

11.4 Placing the Values of the World's Religions

The world's religions are locally collective social organizations which mix cognitive views of the whole of man's place in the scheme of things, prioritization of values, codes of behavior, and the individual's deepest senses of ultimate meaning and purpose. They operate with the entire multiplicity of human concerns, personal and social through transcendent. Our concern is with the relations of the values of any and all religions to the value structures of the individual inner sensibilities as these have been presented here. Tables 11.1(b) and 11.1(c) also identify central values recognized by the world's religions to round the description of human values initiated in Table 3.1.

The values of the world's religions can be conveniently grouped as referring to: God, attitudes of heart, meditative traditions, value-entry, virtue, code, dogma, and mythology. These, in turn, can be related to the domains of value identified in Table 3.1 as follows: God and attitudes of heart (quintessential value), meditative traditions and value-entry (valuative-representation-essence), virtue-[personal, worldly, general] (intrinsic personal primal, intrinsic interpersonal, and sociopersonal values, and valenced truth), and misc code, dogma, and mythology (valenced truth).

A sparse but representative array of values from the various world religions is arranged by this scheme in Table 11.1(b). The values are shown by group in the second column ('Religious Values'); the main corresponding sensibility capacity and domain of value are shown for each entry in the first column. The two right-most columns show the corresponding seats of validation and balancing ('criteria') and typical operational level.

In this table the more fundamental values group toward the valuative-representation-essences. These values are often the most strongly motivating for the deeply religious. Many broadly recognizable values collect in the mixed realm of the sociopersonal values. These have pervasive significant involvement of the intrinsic personal and intrinsic interpersonal values, either explicit or implicit. The more variable values are associated particularly with the realms of code, dogma, and mythology. We see these as valenced thoughts. Table 11.1 (c) offers a fuller sketch of main values of the world's religions. (See, for example, Solomon and Higgins[3].)

This approach brings important values of the world's religions individually into the realm of potential adoption and use by any individual in interbalance with all his other considerations in any particular instance. The approach also serves as a ground structure within which the elements of various religions can be considered and compared. This can reveal: common, even universal, ground; increase of understanding by outsiders; and perhaps some progressive reform where appropriate. In both instances the elemental religious value statements are related to their root type in the generic universal human inner sensibilities. The place of religion and the religious values of the world within the operating value systems of individuals, in or outside of any particular religion, is discussed in Chapter 14.

Endnotes

1. Popper1959
2. Table 11a is summarized from K Nielsen, pg 81-116, v 3 of Edwards1967 and R Norman, pgs 586-591 in Honderich1995.
3. Solomon and Higgins1966, Higgins2001

Table 11.1(b). Placement of values of the world's religions — Part 1.

SENSIBILITY REALM	RELIGIOUS VALUES	CRITERIA / OPERATIONAL LEVEL
	(mythology)	
	-the virgin shall be with child	transcendent
	(miscellaneous dogma)	
	-there is no God but Allah, and Mohammed is his Prophet	Popperian: validation,
valuative representation		synthesis expansion
	(misc code)	growth, universal
	-do not kill, steal,	growth
valenced thought	bear false witness, ...	
	-no adultery	
adaptive integration	VIRTUE:	law, culture, religion
	-justice, temperance, prudence, courage	local group individual seats of
	-stewardship	experience, judgment
SOCIOPERSONAL	-purpose, integrity, dharma	
(mixed	- ...	radially- personal
plastic-intrinsic)		{centric,expansive,transcendent}
biological primals		
	VALUE ENTRY:	
	-the way that can be known is not the constant way	universal
valuative	-treasure righteousness, not world	ultimate individual seats
representation	-enter thine inner chamber	of experience, judgment
essences	-non-attachment	

Table 11.1(b). Placement of values of the world's religions — Part 2.

	MEDITATIVE TRADITION		
	-there is a state..the uncreate..	intrinsic	
	the end of sorrow	valuative	
valuative	-complete surrender to truth	representation	
representation	-another dimension, silence,	essences	transcendent
essences	space		
	-non-attachment, innocence,		
	vulnerability, bliss	meditative ground state	
	- ...		
	-		
	ATTITUDES OF HEART:		
	-preciousness of all life (ahimsa)	intrinsic	
	-faith, hope, charity	valuative	transcendent
	-do not covet	representation	
	-love your neighbor as yourself	essences	
	-forgive, judge not		
- - - - - - - - - - - - -	-resist not evil,	- - - - - - - - - - - - - -	
	love thine enemies		
	-reverence, humility	[peace beyond understanding,	
	-...	ultimate meaning,	
variably		preciousness of all life,	
		infinte sweetness, ultijmate	
obscure-accessible	GOD:	commonality of all life before death,	
	-seek the godhead within you	all-inclusive benevolence, love]	
	-love the lord thy god with all		
	thy heart, soul, mind		

Table 11.1(c). Values of the world's religions.

mohammad
-charity, hospitality
-honesty
-integrity, courage
-eternal struggle of good and bad, religious tolerance
jesus
-return good for evil (love & pray for those who injure you)
-be tolerant, judge not, make yourself perfect
-love others as yourself (do unto them & not, as you would they you)
-seek peace, goodness, mercy; eschew anger, lust, vanity, material treasures
-love the lord thy god with all thy heart, soul, & mind
buddha
-right ethics-{thought, speech, action, livlihood}
-right mindfulness-{.. contemplation, compassion, .. middle path}
-freedom from sensuality, desire, passion, selfishness, anger, hatred, envy
-complete surrender to truth
-deepest inner sublime peace-giving contemplative space
taoism
-flow spontaneously in here-now, wu-wei, fit into environment
-attune inner nature with natural external flows
-avoid artificiality, prescriptive codes
-the way that can be known is not the constant way
-transcendent identification with whole (the way)
confucianism
-inner goodness precedes external action
-reciprical obligations of position, complementarity
-examplism, sage, rituals
-respect of community and ancestors
-individual moral cultivation
judaism
-purity rules
-discipline, sacrifice
-care of family, community, poor
-living engagement with god, holiness
-fear the lord, jehovah, and follow his commandments
hinduism
-spiritual aims-{[worldliness-artha, pleasure-love-kama, purpose-place-dharma, spiritual release-moksha]}
-action-karma-{non-attached service of universal goods/ [child, student, householder, forest-dweller, renunciate]}
-devotion-bukti-{only by love does one know god--supercedes all else}
-wisdom-guina-{see thru illusion-maya / false dualities / all is spirit, brahman(in cosmos)=atman(in individual); homology}
-ultimate sacredness of all life, all being - -
{ahimsa=no harm to: oneself, all things, entire universe}

PART C

ON THE CONTEXTS OF THINGS HUMAN

Chapter 12

A Way of Placing Things

The map of this chapter initiates this book's attempt to characterize the contexts of human living. It places the characterizations of the inner sensibilities from parts A and B within the fundamental view of human placement and human knowing described in Chapter 9. In this, it provides a recognizable ground structure for countering relativism and for systematically approaching a description of human circumstances which is undertaken in the following chapters. The map has several other qualities which render it interesting in its own right.

Chapter 9 has suggested that common human experience produces an inherent generic mapping of the whole of one's surroundings in terms of a nested triad of one's self and inner sensibilities, other people and the whole of humanity, and the surrounding all-inclusive whole of nature, implicitly recognizing each of these as housing distinct forms of energization relative to one's self. The map is outlined in Fig. 12.1. This chapter will discuss its qualities and fill in its inner structure with three levels of headings. Chapter 13 will use it to help characterize the human condition, which in turn will be used in Chapter 15 to help identify the primary contextual influences on human living.

12.1 Qualities of This Map of the Whole of Things

Most fundamentally, the map identifies and directly reflects sources of energization (causation) in our universe in relation to us as individuals[1].

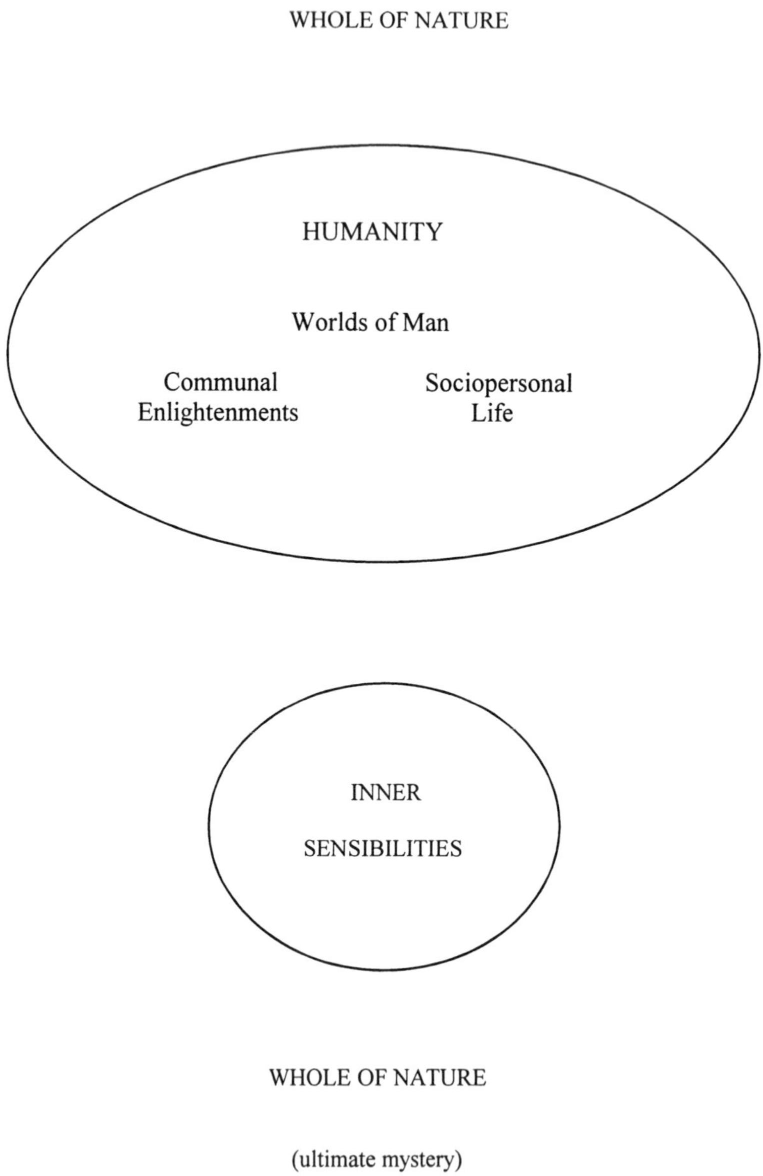

Fig. 12.1 A map of the whole things.

Our partially autonomous human causal power is found solely within our inner sensibilities and self; we are intrinsically like other people but separate, and can readily empathize, interact, and form richly cooperative collectives with them; the forces and behavior of physical nature are different from our autonomous action, beyond our pale. Some animals are remarkably similar to us in some ways, but are never as fully like us or as fully engageable with us as are other people. We will use this characteristic of the map in the following chapter as the basis of a systematic approach to describing the universal human condition in terms of underlying forces, energizations, influences, and constraints.

Secondly, the map is useful as a way of placing all things and ideas in human experience and imagination, as they occur in inner or surrounding reality or are constructed physically or mentally by us. In this it serves as an instructive vehicle for the continuing study, understanding, and cultivation of things and ideas, according to natural familial connectedness arising from their proximities to underlying realms of causation. In this it reveals a natural order within a chaos of information and places a vast amount of conceptual material in easily accessible form.

In this chapter we formally define first- and second-level realms of this map, which will now be referred to as a way of placing things, and indicate a further third-level partitioning. The map can be completed by a final fourth-level partitioning and opened to various significant 'structural pieces' (such as the structures of bodies of knowledge and the like) neither of which is included here.

Thirdly, the mapping displays external reality as we experience it in here-now reality. Most visual scenes of human experience display immediately differentiable manifestations of self, humanity, and the whole of nature. This obvious fact underscores the deeply pervasive reality and rudimentary foundationality of this causational-based mapping.

Fourthly, the map identifies the individual intrinsic viewing point we each occupy from birth to death. It is both uniquely our own in richness of particularity, and universally common with all other humans in its generic structure. In this it identifies the common ground structure of human existence and intercourse. It is not culturally or otherwise relative.

12.2 Realms and Headings of the Map

The interrelations of the map's realms are to be recognized along lines we take as reflective of the fuller natural human process of global arranging. We allow all realms to possibly interpenetrate and overlap as may become appropriate. We take one realm to be in part all-inclusive. We correspondingly allow the definitions of realms to become flexible or fuzzy, even multiple, if need arises. All realms or subrealms contain and retain implicit regions for generalities and for miscellaneous elements. When defining constituent realms or partitions within parent realms, the parent realm always retains room for further additional realms or other items. Labels of all realms are to be taken as broadly inclusive and interpreted in terms of its parent realm. The parent realm of first-level realms is the unrestricted, all-inclusive whole of things. Any given realm may include any and all items of or about: its substance, its context(s), or its influences.

These qualities prevent items from becoming lost 'between the cracks', facilitate ease of access, allow organic growth, and seem more generally kindred and compatible with full substantive and contextual apprehension.

Table 12.1(a) identifies five first-level constitutive operative realms of the whole of things, and five second-level realms for each of these. All of these realms indicate particularly significant unique energizations of human experience within their parent realm, and relate secondarily with other items in the same and other first-order realms. Table 12.1(b) identifies further third-level headings for the all the twenty-five second-level realms.

12.2.1 *First- and second-level realms of the map*

Inner sensibilities in this map is the integrated realm of human neurobiology and mentoexperience indicated by the views described in this book, but also includes all speculations, valid or not, regarding the brain, mind, and personal experience, and inclusive of all functions, structures, properties, and qualities thereof. In the way of placing things, inner sensibilities and self always have reference to a particular experiencing individual--either one's self in one's richness or the generic individual experiencer of the universal human perspective. The second-level realms of the inner sensibilities are its: capacities, functional organization, inner depths, vision and direction, and primary realms of experience. These, in turn, are further indicated by their third-level subrealms in Table 12.1(b). The last three are discussed further in Chapter 15.

Humanity is broadly inclusive of all things of human substance or human origin, especially collective humanity, groups and interpersonal interactions, and other people. This includes all material, adaptive, technological, scholarly, and artistic constructions; all artifacts, skills, vocations, and professions; all individual and collective thoughts and conceptualizations. The scope is best indicated by the realms and other items of Tables 12.1(a) and 12.1(b).

The way of placing things identifies three first-order realms central to humanity: the sociopersonal, the communal enlightenments, and the worlds of man.

The sociopersonal is the realm of inherent human life with particular consideration of collective and interpersonal dimensions. It includes areas of: individual life, people, karma, social structures and protocol, and arenas and intersections. Individual life includes the third-order headings of: generic individual life; history, place, achievement, legacy; business and mechanics of living; public and professional/vocational life; private and social life.

Communal enlightenments includes all human constructions and associations involving the cultivatation and promotion of the higher sensibilities. Paralleling the tripartite structure of the inner sensibilities (biological primals, adaptive personal integrations, and inner valuative representations), the communal enlightenments include three main groups of all constructions which: satisfy the needs of our biological and communal living; effect communal governance and oversight; cultivate the artistic, cognitive, and spiritual capacities of individual and

Table 12.1(a). Constitutive operative realms of the whole of things.

WHOLE of NATURE • sociobehavioral universe • mentoexperiential universe • earth & biological universe • cosmos & physical universe • ultimate mysteries	COMMUNAL ENLIGHTENMENTS • oversight • aesthetics . arts-letters . humanities • understanding-reason-knowing-truth • value-goodness-religion • civilization and communal well-being
INNER SENSIBILITIES • primary realms of experience • vision & direction • inner depths • functional organization • capacities	WORLDS of MAN • places of the world and travel • current living world • history and social theory • world cultures • settings . origins . rudiments
	SOCIOPERSONAL • arenas and intersections • social structures, protocol • karma • people • individual life

communal life. Communal enlightenments is a particularly vital constitutive operational realm of humanity because it relates the infrastructures and highest achievements of our collective worlds to the intrinsic inner sensibilities of each individual human. It is important to recognize that any individual has immediate direct access and authority of personal creation and personal judgment in these realms, although some communal needs value some more than others. The second- and third-level items of the communal enlightenments are shown in Tables 12.1(a) and 12.1(b).

The worlds of man represents primarily all formal or quasi-formal communities or groups (social or special-interest groups, bands, tribes, cities, nations, civilizations). The sociopersonal, communal enlightenments, and worlds of man stand alone as first-level realms within humanity, yet all overlap with each other and with the inner sensibilities of individuals, and any of all these may be considered less centrally, more summarily, and contextually from within any of the others. Humanity also retains space for additional material beyond or overlapping these three central realms. The sociopersonal and communal enlightenments are vital central ingredients in all human collectives, currently culminating in the large democratic civilizations of our times. The second-level realms of the worlds of man are: settings-origins-rudiments; world cultures; history and social theory; current living world; places of the world and travel. Third-level entries for all these, in turn are shown in Table 12.1(b).

The whole of nature covers the same ground as the whole of things but is: more particularized; structured centrally to the 'nature' of common sense and science; and includes the understanding that all particularly human things are also considered separately; therefore, human things, although also fully within the whole of nature, nonetheless are usually treated therein less centrally, more summarily, and within the context of larger objective or extra-human perspectives. The whole of nature includes all of the 'nature' of science and all the substance of science itself, but is not seen here with the reductive views and postures of science. Thus, the whole of nature includes everything —matter, life, consciousness, uncertainties, realities, mysteries, and all qualities and speculations both within or outside of science, reflective of reality or not. The whole of nature is represented in two places in Fig. 12.1 because it is fundamental to human existence in two ways, and because it is all-inclusive and completely ensconces us. The whole of nature includes the earth from which our bodies and brains are constructed and maintained; our inner being and life itself are outgrowths of it and depend upon it. Alternatively, our functional existence constantly engages and interacts with nature through all levels of our sensibilities.

The second-level realms of the whole of nature are: ultimate mysteries, cosmos and physical universe, earth and biological universe, (natural history and) sociobehavioral universe, (living sensibilities and) mentoexperiential universe. These labels represent areas equally recognizable by both science and living human experience. They define distinguishable realms of energization of things human. They structure the whole of things according to the scientific model but allow its easy integration with the whole of human experience and endeavor.

12.2.2 *Comments*

The way of placing things as represented in the two tables of this chapter can be rounded out by the addition of unlimited numbers of 'structural pieces' or 'miscellanea' at any realm of any level. For example, highly useful structural pieces can be constructed readily from professions or academic topics, courses, or areas to enrich many realms of this map. Further, one could further round out the map by adding additional arrays of fourth-level entries for some of the hundred-odd third-level entries shown in Table 12.1(b), beyond the few of these indicated in the table.

Also, one could augment the way of placing things and increase its flexibility by adding a zeroth-level realm to include the philosophy and criticism of such mappings and alternative formulations of first-level realms. Although a best realm can be found in this mapping for almost any particular item, one could define a realm of 'cross-currents' or 'intersections' for marginal elements either within such an added zeroth-level realm, or as a first-order realm in the mapping presented here, or within any stated realm or additional realm.

This way of placing things is useful primarily as an objective and comprehensive way of placing and accessing information. It is particularly appropriate for academic fields and scholarship, or for other objective characterizations of the collective human concerns. Yet, it is not well-suited for characterization of human living. We now will draw on this map in chapter thirteen to approach a description of the human condition, and then use these in Chapter 15 to help map out the contextual energization fields which govern individual human living.

Endnotes

1. The 'whole of nature' is represented twice in Figure 12 and some other places in this work because it shows two important faces to humans: first, it is the ground substance from which we evolved and are ultimately composed and maintained by; second, it is an active external arena which we observe and in which we engage and carry out our affairs.

Table 12.1(b) A way of placing things (third-level headings) — Part 1

INNER SENSIBILITIES

FUNCTIONAL ORGANIZATION
- inner reflective space
- consciousness, autonomy
- inner constructions
- adaptive integrations
- motivation-energization

CAPACITIES
- aesthetic perceptions
- understanding, reason, truth
- value, goodness, badness
- adaptive integrations
- biological primals

VISION & DIRECTION
- intentions . plans
- quest
- purpose . dharma . role
- inner reflective space

INNER DEPTHS
- harmonizations
- transcendence
- body . mind . spirit

WHOLE of NATURE (cont)

SOCIOBEHAVIORAL UNIVERSE
- rudimental human social behavior
- animal social behavior
- neurobiological foundations

MENTOEXPERIENTIAL UNIVERSE
- living sensibilities
- brain & mind sciences
- organic monistic functionalism
- the mentoexperiential
- first principles of awareness

EARTH & BIOLOGICAL UNIVERSE
- outdoors
- phylogenetic tree
- macrobiology, ecology
- anatomy-physiology-medicine
- microbiology
- origin-evolution
- earth

COSMOS and PHYSICAL UNIVERSE
- technology
- quantum mechanics, strings
- lawfulness
- energy-matter-change-motion
- cosmos-space-time

ULTIMATE MYSTERIES
- frontiers
- science
- spirit-metaphysics
- ultimate reality
- God

INNER SENSIBILITIES (cont)

PRIMARY REALMS OF EXPERIENCE
- •communal enlightenments
- •worlds of man
- •human nature. Intersections
- •individual living
- •whole of nature

-inner experiential centers
- •transcendence
- •moods. reflections. reverie
- •fountal streams of good & evil
- •inner depths

- primals
- •death
- •time. change. aging
- •pain. loss
- •motivational complex
- •feelings. emotions. instincts
- •comfort. pleasure. joy. love

Table 12.1(b) A way of placing things (third-level headings) — Part 2

COMMUNAL ENLIGHTENMENTS
OVERSIGHT, ARTS
ARTS, LETTERS, HUMANITIES
-literary arts
-spatiovisual arts
-auditory arts
-somatic & performance arts
-personal arts (life & home)
-types of arts
-art history
-practice & criticism
-aesthetics
-settings.origins. rudiments.milieu
UNDERSTANDING, REASON
-skills, trades, professions
-scholarship
-pure-[philosophy, math, science]
-wisdom, world philosophy
-speculative
-settings.origins.rudiments. milieu
VALUE, GOODNESS, RELIGION
-secular, incidental, popular
-rational
-great religious traditions
-foundations
-settings.origins. rudiments
CIVILIZATION, COMMUNAL WELL- BEING
-human cultures. globalization. world
-government
-oversight & vision
-human rights
-communal functional structures

WORLDS of MAN
PLACES OF THE WORLD
-travel
-world leisure, play, resorts
-cities
-cultures
-national geography, outdoors
CURRENT LIVING WORLD
-contextual energization fields
-events.trends.issues.life styles
-channels of communal life
-institutions
HISTORY & SOCIAL THEORY
-modern democratic society
-law
-government
-philosophy of history
-history
WORLD CULTURES
-general, overview
-individual cultures
-maps of cultures
-peoples
- SETTINGS.RUDIMENTS
- - -

WHOLE of NATURE
PRIMARY EARTHLY ENERGY SYSTEMS
-sky and cosmos
-large issues of the day & spirit
-human activity centers
-human geographics,peoples,cultures
-eras, life forms, ecosystems
-earth, geography, outdoors

SOCIOPERSONAL
ARENAS & INTERSECTIONS
-events. encounters. social
SOCIAL CONVENTIONS
KARMA
PEOPLE
-greats.heroes.human potential
-personal spirituobiological qualities
-friends. people. celebrities
-human diversity
INDIVIDUAL LIFE
-private & social life
-public & vocational life
-mechanics of living, business
-history. place. achi.vement. legacy
-generic individual life
- - -

Chapter 13

The Universal Human Condition

This chapter pulls together a wide range of observations and implications from all the preceding considerations regarding the first principles and qualities of our inner sensibilities and place to produce a sketch of our raw, somewhat muddled, and incomplete intrinsic nature and inherent condition within the whole of things. This topic is central to the attempt of this book to characterize the circumstances of human living. The topic rests upon a vast amount of material which can be approached systematically from the previous material of this book, but can be only scanned here for its most salient points. The approach produces a rough but respectable impressionistic sketch of the characterization it seeks, which I believe does identify the major factors and overall tenor of the universal human condition.

This sketch is presented in four sections as: Fundamentals of the human condition (drawn mostly from the first principles of Chapter 9); human nature (drawn mostly from the previous chapters on the inner sensibilities in Parts A and B); science and nature, and worlds of man (both reflective of the map of chapter twelve). The chapter closes with two broad and more systematic summaries of the universal human condition.

This view of the human condition offers a universally recognizable common ground wherein many various specialized realms of individual and collective being can be seen together within the broader pageant of the human potential. It is a bedrock of our rudimentary individual and collective contextual circumstances within and before ourselves, each other, our communal worlds, and nature. All of the remaining chapters draw on it.

13.1 Fundamentals of the Human Condition

13.1.1 *Individual existence in itself--foundations*

These comments are drawn from the first principles introduced in Chapter 9.

We exist as one with an individually unique (although generically common) realm of conscious experience which we alone see from within, and as a source of partial autonomy with which we each uniquely influence external reality through our body.

We exist within a universe of change, which with differentiation, implies qualities of energization, time and space. We experience space and time as dimensions wherein we localize our moments of existence, and mark these off in regularity of intervals and progression which seem to us consonant with the constraints of external reality.

Energization, and its quality of causation, drives the changes and processes of the universe. We infer its presence underlying the various determinative forms of causation in surrounding physical reality. We infer the action of some of these in and on ourselves, and directly experience our own partial autonomy as a distinct source of causation, comparably operative in our location of existence. We infer this presence in various degrees within some uncertain range of other apparently partially autonomous living creatures.

There is fundamental obscurity and perhaps ultimate mystery of origins and ultimate nature of things and of our own awareness.

There is obscurity of apprehension and knowing because of restrictions of our size, localization, and processes of perception and apprehension. (Our pre-analytic apprehensions are extremely limited; our rational empirical processes are inherently circular and regionally localized; there is circularity in any attempt to root our understanding or awareness in the brain.)

We share a kindredness of qualities of experience, expression, needs, and intercourse with other humans.

Our conscious experience and causal partial autonomy relate us directly to a fundamentally mixed experiential-physical/functional existence.

We feel ourselves as existing in time at an ever-progressing present instant (becoming past) between a fixed (but continually accreting past and a partially open future (becoming present). Nonetheless, we exist reflectively and operationally as though smeared out over epochs of time in a mix of sensory perceptions, reflections, planning, and action.

We exist and operate over a radial range of concerns: personal, intimate, local or large-scale groups, universal, and transcendent.

13.1.2 *Individual existence in terms of surrounding reality — foundations*

To these mostly self-reflective qualities of our human condition, can be added these further characteristics more reflective of our surroundings and relation to them.

We exist and operate within a universe according to the three nested realms of our inner sensibilities and self, humanity (others and collective), and the whole of nature (everything, physicality, other creatures, obscurities and mysteries). These are discriminable on the basis of the distinctness of their manifestations of energy and causation, and their relative kindredness to our own experience and the living of daily life.

The whole seems a mix of consciousness (which we experience directly in ourselves and infer in others) and physical matter and energy. There is a mix of lawfulness and partial constrained freedom, and a mix of partial order and tendency to disorder.

We are biologically dependent on external sustenance for survival, and in this have an inherently imposed foundational necessity to act. We are contingent and mortal. Typical individual lifetimes exhibit a given mean value yet can vary little or much according to genes, life styles, or recognizable external or internal causes.

We, as almost all life forms, are born as one of two genders, and ninety percent of us are born into a corresponding bipartite sexuality, which is necessary for the continuation of the species and pervasively motivating. Sexuality is one of a larger number of deeply ingrained and deeply motivating interpersonal bonds, tensions, and coalitions which ensure, vitalize, and flavor our cooperation in intimate and larger collective groups. Complementarity of relationships wherein weaknesses and strengths of partners complement each other is a relatively pervasive ingredient of human sexually-based partnerships and also of many non-sexual ones.

We experience great aggrandizement of power and the human potential by cooperation in large collective groups. These provide: fulfillment of intrinsic bonds, great practical advantages, great collective constructions and achievements, great enhancement and expansion of delights and gratifications, and great interpersonal turmoil.

Primary realms of experience include such things as pain-loss, time-aging, death, meaning, love, empathy, bonding and support, darker sensibilities (resentment, envy, malice, and so on), understanding, transcendence, purpose, interpersonal rivalry, turmoil, and so on. These are the stuff of literature, poetry, and individual living. Many may entail any mix of the personal, intimate, local, universal, and transcendent.

This listing is representative of primary foundations. The next three sections describe characteristics associated with more external or derivative foundations, first of the inner sensibilities in themselves, and then of external reality.

13.2 Human Nature

Our intrinsic nature seems to revolve around our biological foundations, the implications of our dual structuro-functional plasticity, our inner theater of valuative representation, and the range these give to our partially autonomous human causality. Thus, what we call human nature can be largely characterized according to principles and qualities of the inner sensibilities as described in Parts A and B. These might be represented by the following.

We have genetic prescriptions which carry out or drive action to support our biological life in our earthly environment. Brain centers control and maintain requisite homeostatic chemical equilibria of our body's physiology and drive behaviors to satisfy hunger, thirst, procreation, body temperature, and shelter.

We have a multi-leveled motivational complex containing: primitive sensorimotor reflexes; arousal systems mediating startle, arousal, attention, fight or flight; instinct utilizing sequences of coupled trigger stimuli and fixed action patterns; appetitive systems recognizing incentive stimuli and action-contingent situations; affective systems utilizing value-weighting and significance recognition; and plastic adaptivity and freer higher valuative-representations both utilizing learned plastic inner constructions.

Our higher neural plasticity reflects and fulfills a quintessential open-endedness in our apprehensive mapping, valuing, and operating in our surroundings. At a first level, this allows us to adapt and continually self-correct our understandings and behavior to our better overall well-being in the circumstances of ongoing living. Open-endedness seems closely related to the operations and perhaps origin of both our conscious experience and our partial autonomy.

All these factors--open-endedness, adaptivity, consciousness, autonomy--as virtually all features of our inner sensibilities, are inherently partial, incomplete, and of limited potency. Most of our actions and abilities are a mix of nature, nurture, and partially free original individual construction.

Our deepest theater of existence, our conscious reflections, our deepest senses of meaning, judgment, and truth, our conscious experience, our partial autonomy, our capstone organization of function --all are associated mostly with our highest existential/functional realm, that of inner valuative representation. This realm is characterizable in terms of its capacities of aesthetics, apprehension, and value itself. It is our quintessentially unique human capacity. It serves and modulates over our biological and adaptive capacities, and adds our uniquely human qualities of inner reflection, objectivity, universality, and transcendence to overall experience and control of behavior. The relation of our consciousness to the neurobiological and physical remains obscure. Its mystery may relate to the paradox of apparent partial autonomy in the former to presumed determinism of the latter. The resolution of this paradox would seem to require the denial of even partial autonomy, or an expansion of physical science to place autonomous action and consciousness within or relative to the physical world.

Our overall human causality then is mediated in overriding partial autonomy through the several levels of: removed potentially objective and transcendent reflection, value and judgment, intentionality, sociocultural and environmental adaptivity, biological drive, and biological state-- all manifesting in a complex hierarchy of neurobiological systems. All this places us in a corresponding mix of concern across the personal, intimate, group, universal, and transcendent.

We vary individually along many dimensions, including mode type, abilities, character, temperament, size and scope, as is further indicated in the 'snapshot' a few lines down.

Our plasticity gives us great freedom and range of open-ended development and growth, and leaves us open to particularly tenacious early training, for good or ill.

Our adaptivity operates with a mix of genetic foundations and much plasticity of content; our higher sensibilities--aesthetics, apprehension, and value--often cooperate in the production of internal judgments and constructions.

Our knowing of things and estimates of truth are human qualities, which can be characterized as intrinsic, inherent, and plastic contingent types, according to their underlying neurobiology and range of occurrence. Plastically constructed value-charged thoughts are deeply pervasive in the mores, codes, and behavior patterns of groups and cultures. They permeate our religions, morals, and wisdom.

13.2.1 *Snapshot of human nature*

• multiple inner causality, multiple levels, degrees of freedom:
  o-[originality, mix of mode types, operating level, character-abilities-personality, size.depth.scope]
  o-mix of operating levels = [self, local groups, cultural, universal, transcendent]

  o-mix of mode types-[primals: hedonist ; value: moralist-saint ; apprehension: scholar, aesthetic: artist, adaptive: citizen-leader-diplomat]
  o-degree of originality: [genetic (nature) + training (nurture) + autonomous construction]
- radially self-centered, self-concerned; generally communal; prone to fountal streams of good & evil
- organic existential/functionalism-[creature comforts, plastic adaptive, aesthetic, apprehending, valuative]
- need to act, rooted in life on earth-[contingent, mortal, biological & psychosocial wants, sexual, higher judgments]
- partially open-ended, mixed, constrained, incomplete

13.2.2 *Bald skeletal statement of human nature*

Our skeletal human nature is: open-ended plastic self-corrective constructive functionalism-- across the full mix of inner capacities within the contingencies of biological life on earth, and in some variably-mixed intrinsic-inherent-plastic relationships to other humans (intimate & communal)--all partially consciously and partially autonomously directed from within an inner reflective theater of apprehension-valuation.

13.3 Science and Nature

Nature imposes many essential conditions of humans and science has studied many of these, especially over the last two centuries, and many particularly regarding the physical universe and biological life. This important voluminous material is only scanned here for major features relative to the human condition, but is not complete.

This and the next section use, in part, the medium of lists and tables to convey a comprehensive overall impressionistic image of a large body of complex and disparate material which may express any of several facets or dimensions of the material, without engaging or dealing with its substructure in a rigorous fashion. The reader is asked to see these as analogous to a cubist painting which tries to convey a multiplicity of perspectives within the constraints of a flat two-dimensional surface, and in so doing to illustrate essential features of the various views in multiple juxtaposition. The justification and reward is the relatively systematic summaries of the universal human condition given in the last section of the chapter.

13.3.1 *Obscurities and mysteries*

There is significant deep obscurity and perhaps ultimate mystery in a number of places, including:

- foundations and origins
- meanings (except as we find them in life itself or the realm of inner reflections)
- the 'stuff' (matter-energy) and lawfulness of the physical universe
- the wave-particle duality & uncertainties of microphysicality, and strangeness of the cosmos as revealed by twentieth-century physics
- the extreme complex intricacy of the macromolecular chemical organization of life
- the substantive relationship of conscious experience with the brain and neurobiology and the physical
- the apparent autonomy of the human consciousness within otherwise determinative physical law

13.3.2 *The physical universe*

- 'stuff' (matter, energy)
- pervasive physical lawfulness:
  - o-conservation (matter, energy, momentum) / interchangeability-transformation
  - o-[energy-force / motion-heat-combination-chemistry]
  - o-order / direction to disorder (entropy)
- primordial microuniverse: quantum mechanical duality and uncertainty; apparent frenetic spontaneity of origin; strings and quantum theory; deterministic paths of probability distributions
- apparent cosmic strangeness: temporal relativity, curved space-time, alienness of texture, ambient environment, scale, speed limit

- subject to physical forces, energetics, chemistry as matter.
- big bang, earth, planets & sun, stars, asteroids-dust, galaxies, clusters, ...
- constrained to earth by gravity, biological needs, scale, energy needs (life not yet discovered elsewhere)
- restricted to narrow shell (on land) by ambient conditions
- subject to daily, monthly, annual, longer cycles
- buffeted by forces, stressed by temperature changes, radiation, storms, weather.
- smallness before the scale of the earth and the universe.

13.3.3 *The biological universe*

- chemical organization in macromolecules--proteins, DNA
- creation of order
- chemico-organic complexity
- evolutionary long-term plasticity and selection of genetic prescription
- habitats and ecosystems

13.3.4 *The sociobehavioral universe*

- multiple intrinsic placement and radiality
- bipartite sexuality; intrinsic bonds, tensions, and coalescions; complementarity
- instinctual systems with preprogrammed fixed action patterns, archetypes, and trigger stimuli
- adaptive processes of plastic associations and representations

13.3.5 *Conscious experience and apprehension*

- a realm of universal human pre-analytic rudimentary apprehensions of our nature and surroundings indicated here as foundationo-rational: conscious awareness itself, mentoexperiential coherence, higher intelligence, partial autonomy, existence, energization, differentiation, inner sensibility, organic monistic functioning, self, other people, separative intrinsic placement, humanity, surrounding

external nature, sensation, internal reflection, limited understanding
- inner valuative representations of aesthetics-beauty, apprehension-truth, value-goodness-badness
- prone to variably rare powerfully moving moments of deeply impressive peak experiences of exhilaration, joy, liberation, expansion, benevolence, inclusion, love, meaning, purpose, place
- intimate but obscure relationship with neurobiology

13.4 Worlds of Man

This section draws further inferences regarding the inherent human condition which are more reflective of our current lives and our accumulated human knowledge, understandings, civilizations, and material constructions and artifacts. No end of such inferences could be drawn. This lists salient central features, but is certainly not exhaustive.

Human families, bands, tribes, societies, nations, cultures, and civilizations are higher cooperative worlds of man, whose collective codes and policies have played a major role in most human living throughout history.

Cultures and their cohesions are rooted in their constituent individuals. Individuals provide both the ultimate energy, conceptual constructions, and direction for such cultures, and their ultimate seats of knowing and judgment. Cultures are dependent on individual labor, direction, creativity, ultimate seats of judgment, and cooperation.

Sexuality is one of a larger number of deeply ingrained and deeply motivated interpersonal bonds, tensions, and coalitions which ensure, vitalize, and flavor our cooperation in intimate and larger collective groups. Complementarity (of weaknesses and strengths) characterize many sexually-based partnerships and other types as well.

Individuals experience great aggrandizement of the human potential and power by cooperation in large groups and collectives. This includes: fulfillment of intrinsic bonds, great practical advantage, great enhancement and expansion of delights and gratifications, and great interpersonal turmoil.

Individual humans typically find themselves in need of position in or with respect to collective humanity, what can be called ‘a place in the sun’.

Cultures seem to operate with collective interests and themes and to have operated historically with collective inertia and momentum with time periods longer than individual lives.

Cultures acts as pressure cookers wherein volatile energy of interpersonal cooperation and dynamics channels into useful productivity in synergisms well beyond what the individuals working separately could produce. Entirely new and unexpected things emerge.

Cultures accumulate knowledge and productive know-how, with ever-increasing skill for common human needs and wants.

Cultures have been directed and managed by various forms of government, controlled by dictators or law, maintained ultimately by power structures and police, and protected and served externally by armies. Free speech counters with feedback for self-corrective control.

Negatively, cultures produce self-serving cultural doctrinal conditioning, and strong social pressures to support local mores, conventions, prejudices, and all too often, local hatreds and bigotry.

Cultures have often been domineering over individual interests, wants, and sensibilities.

More neutrally, individuals living in cultures generally experience:

- needs of place-position-roles--a place in the sun (this usually draws individuals into a situational homeostasis.)
- many exigencies and pressures of practical individual life
- synergy, kindredness, bipolar volatility of communality, and sociopersonal interaction, including rewarding or disturbing karmic continuities
- a dozen or so fountal streams of good and evil generally pervasive through most things human can be identified in the universal human inner sensibilities across the quality of radiality between the poles of empathy and malice, and more broadly between the inner and outer reaches of the inner reflective space and outer plastic constructions of the higher valuative sensibilities (see Chapter 14)

At root, cultures can be readily seen to be constructed in the terms of the functional capacities of the inner sensibilities of their individuals.

Their primary concerns, their functional capacities, their institutions, their entertainments, their daily lives and operations--all are direct reflections of the individual's capacities of biological primals, adaptive integrations, and inner valuative representation dimensions of aesthetics-beauty, cognition- truth, and value-goodness.

Cultural institutions which embody explicit concern with--biological and situational well-being of all individuals, overall governance and progress of the culture as a whole, and with the cultivation of the arts, scholarship and knowledge, and free spiritual values-- constitute a recognizable realm which may be labeled the communal enlightenments. The collective cultural support and cultivation of these is what this work recognizes as a most advanced mark of civilization.

Cultures in themselves also reflect the general qualities of these individual inner sensibilities. Cultures are plastic, adaptive, ultimately open-ended, and self-corrective. They are inherently mixed, partial, incomplete, imperfect, and operate radially with a general mix of concerns of self, intimates, groups, universals, and transcendentals.

Our fundamental plasticity and the vast accumulated knowledge, skills, tools, technology, material goods, and vital human services of all kinds provide corresponding vast opportunity and degrees of freedom to most people in the west and many throughout the world today. This instrinsic human plasticity characterizes all peoples of the world.

13.5 Broad Summary of the Human Condition

The following broad outline casts the human condition in a representative natural perspective. Fuller sketches follow in Tables 13.1.

- We exist within an intrinsic viewing point in conscious awareness and partial autonomy of obscure origins and margins, partially characterizable according to a few first principles such as intrinsic placement, mind-body obscurity, and organic bipartite functionalism.
- We are rooted in biological life on earth. In this we are contingent,

needful, mortal, and have a foundational necessity to act. Beyond this, we are an uncertain mix of the physical, biological, and consciousness, and set within larger surroundings of obscure origins and margins.

- We are a mix of capabilities and powers, limitations, incompleteness, constraints, and contingencies. We are mixed and partial in all qualities; multi-leveled and radial in interest and purpose across concerns of self, intimates, groups, universals, and transcendentals. We are driven by a multiple internal human causality (including motivation, drive, homeostasis, regulation, adaptation, intention, volition, preference, value, objective inner reflection, goodness, truth, beauty, justice, and many others) and given rare moments of deeper and sometimes ultimate senses of meaning, purpose, and direction.
- Our most significant qualities are: our plasticity and open-endedness—what happens next depends partially on us; our robustness of adaptive functioning, learning, self-correction, and accumulative growth within our natural earthly environment; and our abilities of partially free original reflection, creation, and construction.
- Our quintessence is our partially free partial conscious awareness and its general meandering quest through open-ended reflective frontiers of 'what is this?', 'what is true?', and 'what is good?', according to our intrinsic structuro-functional capacities for constructive representational reflection--aesthetic perception, apprehension, and value.
- We are variably and partially prone to volatile bipolar interpersonal bonds, tensions, and coalescions, and to the formation of cooperative communal cultures. These provide enhancement and fulfillment of practical needs and sensibilities, great aggrandizement of all human capabilities, and a rich milieu of socio-interpersonal relations. In our collectives, cultures, and civilizations, all human existence, awareness, and ultimate seats of judgment and productivity remain within individual humans according to the intrinsic separative multiple placement of our inner sensibilities in individuals.

Table 13.1(a) collects what this work sees as the most significant of the entirety of qualities of the human condition, arranged according to five readily recognizable and approachable dimensions. This table compresses an exceedingly large body of complex material into a rough, but instructive and relatively complete, succinct image of the human

Table 13.1(a). The human condition in five compound tenets — Part 1.

intrinsic placement

-multiple separative intrinsic placement of individual existence, awareness, and partial autonomy
  o-we are primordially individual:
  o-all individual & group experience, judgment, action, originates, registers in awareness of individuals

-inner reflective space & theater, imagination, representations, valuations, judgments
  {comfort-want, pleasure-pain, vitality, joie-de-vivre, feelings, aesthetics, value, good & evil, apprehension, understanding, fear, anger, malice, brutality, benevolence, love, peace, meaning, purpose, mystical intimation}]

-localized within larger obscure inner & outer realms (inner sensibilities & whole of nature)

essential nature

-partially conscious, partially autonomous, partial director of personal action, individually unique
-imaginative, creative, questing
-plastic, self-correcting, quintessentially open-ended
-unique species on earth ~{rich inner valuations, robust creative constructive capacities, questing open-endedness}

practical nature

-primal necessity to act
-rooted in biological life on earth: contingent-mortal-doomed-wanting-naked-vulnerable
-full multileveled motivational complex
-robustly functional/existential organization, adaptive, constructive
-full sensibilities-{biological primals, [aesthetics, understanding-reason, value-goodness], adaptive integrations}
-partial, mixed, incomplete, imperfect, limited, constrained, contained
-subject to genetic prescription (generic & particular), and to training & conditioning by early surroundings
-often: excessively selfish, envious, abusive, cruel, violent; sometimes evil

Table 13.1(a). The human condition in five compound tenets — Part 2.

relational nature

-radial-{self-centric, expansive, objective, transcendent }
-partially autonomous, partially independent, partially collective-communal
-prone to form collective organic groups, from small bands, tribes, to large states, nations, civilizations
-needful of 'a place in sun'
-all individual & communal human experience reflect the qualities of the individual sensibilities:
-interpersonal bonds, tensions, coalescions:
    o-intimate; (partially:) o-social , o-communal-{tribes, states, civilizations, global }
    o-complementarity of masculine-feminine desire, strengths & weaknesses
    o-primal empathy, love, evil
-all interpersonal relations (intimate, social, communal) prone to turbulence-karma-violence-war
-higher enlightenments of aesthetics-cognition-value endemic to human experience (individual & communal)

origins, obscurity, place

-vision unsure: filtered messages at a point in a vast whole--in sensibilities, to sensibilities, by sensibilities
-givens of perception: energization, motion, change, time / discrimination, multiplicity, space, diversity
-existing always:
    o-centered within one's own body, and radially sensible from this point
    o-at a fluid forward-moving intersection of fixed past & partially open future
    o-what happens next depends partially on us
    o-at a livable base of circumstantial being, on verge of partial change
    o-at an instant metaphysically, yet distributed over time in almost all brain-mind-action
-existing mentoexperientially within:
    o-a physical universe of obscure origin, lawfully self-relevant, as if turned on then abandoned to its own prescriptive causes, heading towards a cold uniform stillness by inexorable physical law
    o-self-relevant biological & mentoexperiential realms of mysterious singularity of origin & energy

Table 13.1(b) The human condition by the universal human perspective — Part 1.

relation to worlds of man

-higher enlightenments of aesthetics-cognition-value endemic to human experience (individual & communal)
-all interpersonal relations (intimate, social, communal) prone to turbulence-karma-violence-war
-often: excessively selfish, envious, abusive, cruel, violent; sometimes evil
-all individual & communal human experience reflect the qualities of the individual sensibilities:
-prone to form collective organic groups, from small bands, tribes, to large states, nations, civilizations
-needful of 'a place in sun'
-partially autonomous, partially independent, partially collective-communal
-interpersonal bonds, tensions, coalescions:
o-intimate; (partially:) o-social , o-communal-[tribes, states, civilizations, global]
o-complementarity of masculine-feminine desire, strengths & weaknesses
o-primal empathy, love, evil

relation to whole of nature

-unique species on earth ~{rich inner valuations, robust creative constructive capacities, questing open-endedness}
-robustly functional/existential organization, adaptive, constructive
-primal necessity to act
-rooted in biological life on earth--contingent-mortal-doomed-wanting-naked-vulnerable
-existing mentoexperientially within:
o-a physical universe of obscure origin, lawfully self-relevant, as if turned on then abandoned to its own prescriptive causes, heading towards a cold uniform stillness by inexorable physical law
o-self-relevant biological & mentoexperiential realms of mysterious singularity of origin & energy

Table 13.1(b). The human condition by the universal human perspective — Part 2.

relation to entire whole of Things

-what happens next depends partially on us
-partial, mixed, incomplete, imperfect, limited, constrained, contained
-subject to genetic prescription (generic & particular), and to training & conditioning by early surroundings
-localized within larger obscure inner & outer realms (inner sensibilities & whole of nature)
-multiple separative intrinsic placement of individual existence, awareness, and partial autonomy
-all individual & group experience, judgment, action, originates, registers in awareness of individuals

inner nature in itself

-partially conscious, partially autonomous, partial director of personal action, individually unique
-imaginative, creative, questing
-plastic, self-correcting, quintessentially open-ended
-full sensibilities-{biological primals, [aesthetics, understanding-reason, value-goodness], adaptive integrations}
-vision unsure: filtered messages at a point in a vast whole--in sensibilities, to sensibilities, by sensibilities
-givens of perception: energization, motion, change, time / discrimination, multiplicity, space, diversity
-full multileveled motivational complex
-radial-{self-centric, expansive, objective, transcendent }
-existing always:
  o-centered within one's own body, and radially sensible from this point
  o-at a fluid forward-moving intersection of fixed past & partially open future
  o-at a livable base of circumstantial being, on verge of partial change
  o-at an instant metaphysically, yet distributed over time in almost all brain-mind-action
-inner reflective space & theater, imagination, representations, valuations, judgments
  {comfort-want, pleasure-pain, vitality, joie-de-vivre, feelings, aesthetics, value, good & evil, apprehension understanding, fear, anger, malice, brutality, benevolence, love, peace, meaning, purpose, mystical intimations}
o-we are primordially individual

condition as I see it and wish to convey it in this work.

Table 13.1(b) rearranges the entries of Table 13.1(a) according to human nature and one's relation to each of the three realms of the universal human perspective: one's inner sensibilities, humanity, and the whole of nature. These constitute the three independent centers of organization and energization within which one lives and must interact. In this sense each category in this table reflects, houses, and partially indicates the energizing source of its label.

Chapter 14

Fountal Streams of Good and Evil

This chapter identifies a dozen fountal streams of good and evil and their ground structure in the inner sensibilities, and then relates these specifically to an opposing process theory of empathy and malice as a generalization of the attachment model of human bonding. It then discusses: the problem of karma, the scope and pith of the good life and morality, the place and effect of morality in human evolution, and offer final reflections. The approach is to ground these discussions within the full inner sensibilities and human contexts considered in the previous pages of this work.

14.1 Fountal Streams of Good and Evil

Figure 14.1 identifies a dozen essential streams of the good and bad which permeate our ongoing individual and communal lives, and largely define our senses of the good life and of morality. Their recognition here helps identify the tangible focus of this problem area.

The figure also shows the ground structure of these streams of influence within the human inner sensibilities themselves across the spectrum of intrinsic, mixed, and plastic values identified in chapter three. This identification brackets the various streams, indicates their character, tractability or tenacity, and suggests approaches to their cultivation, amelioration, and management.

The ground structure consists of a broad center region determined across biological primals and adaptive integrations containing two extreme poles of empathy and malice (evil), and two surrounding branches of inner valuative representations (aesthetics-balance, understanding, goodness, and justice).

The center region involves intrinsic, intrinsic-inherent, and mixed sociopersonal values, reaching for the most part across self-other radiality between the two extreme poles of empathy and evil. The reflective valuations, both housing especially the valuative dimensions of aesthetics, understanding and truth, goodness, and justice, partition into a deepest path of intrinsic inner reflective awareness and a more external path of outward manifestations. Fountal streams of the first path exist most directly in inner reflective space itself and here-now experience, and are generally heavily extra-rational and non-language based. The outward path is externally portable, communally available, open-ended, and self-corrective across the same valuative dimensions. This second stream is more rationally framed and expressible in language more outside of immediate experience in itself.

In rough shorthand the fountal streams reflect five fundamental internal influences: inner reflective sensibility in itself, empathy, radiality, evil, and valenced thoughts. Empathy and evil emerge in some obscure mixes of inner reflective sensibility, intrinsic values, and plastic adaptations, and are seen here as the extreme poles of self-other feeling. Empathy seems to manifest more closely to reflective sensibility and biological primals, whereas most forms of badness and evil seem to manifest more in association with adaptive integrations, radiality, and normalish self-centered drives and frustrations. In this, empathy may be more primal and evil more secondary. However, at least one form of evil discussed a few paragraphs down seems to arise from the failure of empathy to develop in non-nurtured childhoods.

The individual fountal streams themselves are largely self-explanatory. Badness is a portion of value. It may be primal in itself (pain, sorrow, intrinsically noxious brain circuits or chemical states), or it may manifest simply as a negative aspect of goodness, such as its absence or frustration. Bad and good sometimes meld into a whole, but mostly lurk separately within human affairs as adverse colorings, say various intensities and shades of two distinct hues. Most fountal streams

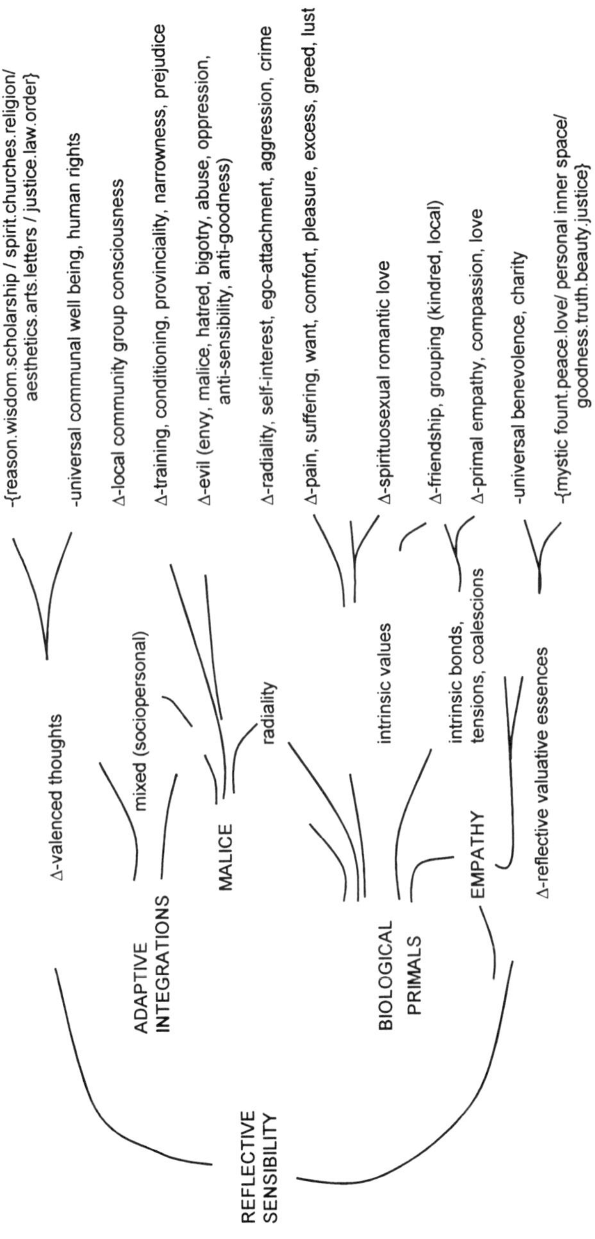

Fig. 14.1 Fountal streams of good and evil with ground structure.

may spawn both good and bad influences. Most fountal streams reflect a mix of the various sources, roughly indicated by their positions in Figure 14.1.

Evil is taken here as a particularly vile subset of badness--quintessentially a willful negative anti-pole of goodness, the qualities and action of a malicious spirit, including: envy, malice, abuse, destruction of the good, beautiful, true, or vital, an anti-sensibility, anti-goodness, anti-hero, anti-saint, anti-sacred. Evil is thus a portion of the primal dimension of value. It seems a mostly human phenomenon: virtually all the evil we encounter is human-made (or in the hands of humans). However, apparently hatred-driven behavior has been described in hyenas, lions, elephants, and chimpanzees.

The fountal streams manifest as inner feelings, attitudes, mental representations or judgments, intentions, propensities to behavior, or behavior, often in implied association with externals and most often with other people, or things associated with other people. Some individual streams may include good, bad, or neutral influences in part dependent on degree, balance, external circumstances, or cultural or group standards. Others may be similarly more or less remarkably good or bad. Often the good and bad are close companions, especially in the flurried mix of worldly activities and confluent interests. They can be often difficult to know clearly, except in purity. Many judgments of good-bad depend on the balance or misbalance of radiality between self and others.

These fountal streams are seen here as the central manifestations of goodness, badness, and evil in individual and collective human life. This structuring of their ground structure and nature provides an indication of how they might be recognized and managed constructively both individually and collectively.

14.2 Empathy and Neural Development

The interplay of the neurobiological with the emotional, valuative, interpersonal, moral, criminal, and evil in the human inner sensibilities is well illustrated by recent research which identifies a clear neurobiological syndrome in right brain function regarding human primal empathy[1]. This syndrome is associated with the neglect, abuse, or trauma of children during a critical period of brain development before the age of two. The system involved is emotional-related neural networks of the right anterior limbic prefrontal area (ventromedial cingulate gyrus and surrounds) and its cortical and subcortical interconnections. It is possible to identify the syndrome by brain imaging techniques (fMRI, PET). Its behavioral manifestation is a lack of the capacity to feel affection or bonds with other people. It is possible, though demanding, to partially correct the syndrome with behavioral training prior to early adolescence, but not thereafter. Many persons suffer lasting emotional disorders of varying severity from this deficiency syndrome. Its most brutal manifestation is in serial killers a remarkably high percentage of whom are victims of this syndrome.

This 'attachment' view identifies a natural ground structure of compassion and bonding, and its tragic pathology in deprived circumstances. The major thrust of the attachment research is, however, positive. In clearly defining a clinical syndrome it brings immediate suggestions for at least both partial prevention at its earliest stages of inception and partial treatment during its later preadolescent school-years. Efforts at comprehensive scanning for potential victims (including possible future serial killers) and intercessions at both these stages in our communities and schools as is done for various infectious diseases can be considered.

A separate finding is that empathy is associated with maturation of the frontal areas, and that full maturation of empathy is delayed in male humans until early twenties. Perhaps this has genetic Darwinian value, giving the human species a genetically programmed warrior class[2].

This attachment model defines a center of primal empathy within the long-recognized emotionally-related complex of obscure diffuse recurrent neural circuitry involving the hippocampus, amygdala, hypothalamus, stria terminalus, and brainstem reticular formation as initially formulated as the classic 'Papez loop'[3]. In such diffuse neural systems a 'center' is a useful but not totally accurate picture of neurobiological operative functioning. (Perhaps 'region' would be slightly more accurate than 'center'.) Earlier classic studies identified centers for rage (amygdala), pleasure (lateral hypothalamus), and noxiousness (reticular formation) in this larger emotional system of lower animals like cats. Much of the system seems to be involved with consummatory appetitive behavior in mammals (for example, carnivores). Even the higher ends of emotional circuitry are probably engaged in some larger systemic organization. In humans this probably engages with higher plastic integrations and inner constructions. The overall psychopathology of serial killers may engage this larger unknown systemic organization as well as the bonding deficiency itself.

This work's dual-process view models the actions of good and evil in terms of a bipolar foundation of empathy and malice within the inner sensibilities and shows a corresponding interpretation of the full range of observed good and evil in human life in terms of these influences in the brain (Fig.14.1). The identification of the empathy center described by Schore and others maps nicely into our model. To round out this picture of good and evil, it is necessary to find a 'center' for malice (which may be quite a task involving diffuse interconnections). It is interesting in this regard that the serial killers seem to be unfeeling and driven to psychopathic behavior rather than malicious. Is an enraged hungry bear evil? Is an emotionally-crippled driven serial killer with aberrant brain connections evil? We think these are abhorrent, vile, but, in the absence of malice, perhaps not evil--the bear perhaps closer to evil than the serial killer.

There may be some mutually inhibiting coupling between the empathy and the hypothesized malice centers. The nervous system utilizes pairs of mutually-inhibiting bipolar opposites widely for basic motor controls (basal locomotion from insects through humans, spinal and brain stem control of muscular action, and brain stem control of involuntary breathing), perhaps between pleasure and pain, joy and sorrow, and perhaps, as we suggest here, between good and evil. Such a malice center may likely operate in association with psychological states such as frustration working through natural motivational systems in common karmic interactions of life. Similarly envy is often a player in malice whose neurobiological grounding and associations are unknown.

14.3 Karma

A particularly pernicious level of the self-other question (discussed below) entails the reverberating effects of good or noxious interpersonal psychological patterns which are frequently associated with actions in the so-called law of karma. The observation is that often the internal psychology and ultimate inner intentions (for good or bad) associated in the actor with a given action or the action itself generates corresponding internal psychological patterns in the inner sensibilities of others, who in turn pass on this residue to yet others possibly including the sources from which it began in potentially unending reverberation. A number of interesting half-truths are associated with the law of karma: 'What goes around comes around.' 'One reaps what one sews.' 'One's life is determined by one's character'. More generally, worlds of man are perennially infected with good and bad karmic residues, which latter very few can entirely avoid even with requisite inner cultivation.

Relatedly, it seems the case that especially those who are different, but also those who excel or those who seem unusually good beyond local standards, find themselves in receipt of much gratuitous hostility. Envy seems an all too common human phenomenon especially in unfulfilled, stifled, or proud people. It's as if many people are somewhat like crabs who pull back down any of their number who begin to escape up the sides of a pit containing them all. Quite remarkable. More generally, there are those who aspire to high ideals of achievement or goodness (with varying degrees of attainment), others who like to see such ideals in our institutions and in some others but don't feel inclined to so commit themselves, others who dislike those who appear to represent but not attain high achievement or ideals, and yet others who dislike those who do attain high ideals or the ideals themselves or both. Interesting associated half-truths are: 'We pay for who we are--special people have special things to pay for.' 'We die of our strengths'--perhaps from extending them too far. 'If everybody hates you, you know you are doing something right.'

On balance all around, in the world of karma: 'These minor matters require major thinking.' At the personal level these seemingly minor and usually petty karmic perturbations can seriously mitigate the pleasure of life and lead to abrasion or violence. They are the underlying substratum of fundamental tenets in most major religions. On the cultural level, consequences of increasing rather than decreasing hostilities can be catastrophic.

14.4 The Good Life and Morality — Scope and Contexts

The good life, morality, religion, and ethics are all ultimately answerable to goodness itself, which is the underlying source of all these and of all significance in human life, and which may be taken as tantamount, coequal, with the dimension of human value (badness being seen as its counter pole).

This field of topics ranges across a number of dimensions including: the secular to sacred, the rational to fuller extrarational sensibilities, behavior to inner attitudes and mystic intimations, and the radial dimension of personal-intimate-group-culture-universal-transcendent. Indeed a central factor throughout is the relation between self and others. This field of topics presents many difficulties of foundation as indicated in the first paragraphs of the next section below.

Both religion and ethics have important substantive content which may be taken as contributory to our concern, but both are peripheral to the fundamental human-centered approach taken here. This work recognizes ethics in passing as the rational western philosophy of good actions and their motives and ends, and has commented on this area in chapter eleven. This chapter seeks both a more direct founding within the existential-functional ground structure of the brain and conscious experience as outlined in this work than is found in ethics, and a wider openness to the full extra-rational foundations in human sensibilities. This work also recognizes religions in passing as views of the whole of things and man's good and moral life in it according to beliefs and tenets beyond reason, and attributed to sacred revelations, and comments on these in chapters eleven and sixteen. Here also the work seeks a more direct grounding of both morality and the good life within the brain and the human condition as any and all can directly experience it. Also, it is deemed desirable to clearly distinguish the good life from morality, even in recognizing their mutual interrelation and possible total merging. The speculations of Joseph Campbell on the symbolic significance of the world's myths might be helpful in the attempt to identify possible intrinsic-inherent archetypical manifestations of good and evil in human sensibilities (Chapter 10). In this chapter our bottom-up view of these comes toward common ground with Campbell's top-down view, although it is beyond our purview to pursue this[4].

We now focus on the good life, morality, and goodness. Following western philosophy, the good life is taken in the sense of a personally fulfilling life[5]. Such a good life is seen here as the optimal living of an individual life given one's inner sensibilities, human nature, and the human condition, in relation to one's surrounding circumstances, and the widely-free human potential. It is seen here as related essentially to one's own private inner concerns, judgments, happiness, contentment, achievements, and personal freedoms and flourishing, as judged within and by one's own inner sensibilities.

Morality is intimately related with the good life and with the fundamental qualities of goodness. At its most superficial level morality is defined as good behavior. Morality thus inherently entails more explicit senses of other persons, and more external communal pooling than asked of the good life, and a closer relation to formulations in communicable language, which the good life need not necessarily ask of itself. The good life might say 'if it feels good, do it'. Morality asks more.

A central dimension of all these topics is the relation of one's self with others--the primal radiality of individual existence. The interweaving of the good life and morality is a generalization of this central self-other question. Also and more practically, self-other relations are the central glue which sustains all interpersonal activities and all social order, including all bands, tribes, states, and civilizations. These latter concerns engage many secular interests often quite distinct from those of morality with its relations to goodness, thus determining at least a multiplexed alliance between the secular and the good within our cultures as well as in ourselves.

More fully meaningful human morality, however, is deeper than behavior in itself. Attitudes of heart are also central. Primal empathy or compassion is the intrinsic attitude of heart in human self-other morality. The passive condonation of unprovoked willful violent abuse seems morally wrong, for example, even if not legally so. More generally, this work takes morality as good behavior reflective of good attitudes of heart in turn reflective of the entirety of the goodness dimension which includes and interbalances from simplest individual joys or pleasures, through radial self-other relations, to the ultimate founts of mystic experience, peace, and meaning, to universal love and benevolence, to transcendent self-sacrifice or universal service.

Fuller sensitivity to individual inner sensibilities leads to moral inclinations regarding both one's own inner life, with its varied qualities of choice and outer direction, and empathy with the full existentiality of others. At the deepest levels morality is heavily directed by one's inner relation to the dimension of goodness in itself and in this by one's sense of relationship to God.

Morality is ultimately a matter of individual judgment and conscience because qualities of the good life or morality live or die, thrive or wither, within our individual personal spaces of apprehension and judgment, our partial autonomy, and our quintessential open-endedness, and because a unique rich diversity of particulars surround any individual's radiality of circumstances including wide differences in abilities, vision, commitments, activities, plans, inner weightings, and so on. The ultimate registration of all things is for each of us in our individual inner seats of judgment in primordial aloneness in being, before death, eternity, and our sense of God.

Both the good life and morality are ultimately answerable to goodness as seen within one's intrinsic seats of judgment. Beyond this, it is not clear how good any given individual should be expected to be, or, indeed, how good any particular individual should want to be, except perhaps to that person. There are surely inner callings of many shades felt by many; many feel the deepest and highest callings of divine communion and devotion; others may like the ideal and model of the higher virtue or nobility within our collective presence, but prefer to live a more normal human life themselves; others may find fulfillment within creative achievement, totally worldly lives, or the raw vitality of adventure, pleasure, and joy. Ultimately questions like 'how good should a person be?', 'how good it is reasonable to expect others to be'?, or 'how good should I be?' are better left to individual inner valuations and conscience, except for certain essentials regarding the safety and security of individuals and the integrity of communal functioning.

Primal empathy, enhanced inner sensibility, and communal cohesion demand a proscription of abuse of people and living creatures, and also decry the destruction of useful, good, beautiful, true, and just things,

A minimum morality might entail 'carry your own weight plus a little' and 'do not abuse nor hurt others', while a maximum morality might see the fullest human fulfillment in the complete forsaking of one's own good and self for the universal or transcendent, or the good of others. Between such extremes there are wide variations in the range, depth, and content of morality recognized by individual humans, and also differences of emphasis and some disagreement across the various religions of the world.

'Carry your own weight plus a little', can be seen as the just necessity of the universal conditions of our common intrinsic placement, radiality, and the necessity to act; 'avoid abuse or injury of all humans, and, whenever feasible, of all living creatures', can be seen as necessarily following from inner sensitivity, primal empathy, kindred affection, and plastic practical requirements of communal living.

Goodness, truth, beauty, justice are often graded, rather than absolute. Thus, high purity may be approached when pursued locally, but necessarily entails some sulliedness when engaged broadly,

Both 'the good life' and 'morality' are rational guides to living. Living with good and bad in human circumstances in fact engages a full array of sensibilities including both rational and significant extrarational components. They come into play in richly full and diverse choice points with important idiosyncratic nuances of intention, confluence, and circumstance and involve multiple conflowing half-truths (often reflecting valenced thoughts) all of which must be balanced out according to local circumstances. Moreover, balancings may be continually reevaluated according to growth and change.

Often in practice, particularly among the mixed contingencies of worldly life and the myriad half-truths of valenced thoughts, the good and bad are often very close companions; it seems only the pure can approach avoidance of all taint of badness, and only very few of these fully attain it.

14.5 The Good Life and Morality — Pith

The basic requests of morality seem quite simple in outline, but highly complex and volatile in practice. The key seems simply to hold and show respect, empathy, compassion, and non-harm to others, and to balance self-interest with self-sacrifice for others and universal or transcendent concerns.

This is the basis, yet difficulties emerge because value choices arise within muddled middles of multiplexed situations which require the balancing of many values across multiple persons and relationships, and often force central existential choices from individuals. These, in turn bring to the front a host of uncertainties and personal differences of interest and perspective which can render even apparently simple situations difficult, controversial, divisive, emotional, and volatile, including: lack of a universally recognizable view of the origins, nature, structure, and relations of human value; obscurity of origin and purpose of the cosmos, humanity, life, or awareness itself; diversity, richness, subtlety, obscurity of individual genetics, experience, circumstances, and nature; dependence of these things on culture and individual training and conditioning; a sense that overstatement, perhaps any prescription, trespasses into private realms of individual inner reflective space and judgments given by intrinsic placement; a sense that such essential judgments are, according to our intrinsic functional organization, better left largely new, open, ever-freshly respondent to current circumstances, individual circumstances, recent development of surrounding culture and environment; similarly, such judgments may seem sometimes better left to the learning of self-corrective development; ultimately value is outside of reason, and sometimes beyond reason; humans sometimes differ widely about values in themselves.

The pursuit of a personal fulfilling good life seems and is recognized in the west as an intrinsic-inherent quality and right of any human being. There is a remarkable centrality of the few primary components of the individual inner sensibilities (biological primals, adaptive integrations, beauty, truth, goodness, justice) to our individual vocations, our communal functional structures and cultural institutions (communal enlightenments), and the make-up of our popular conceptions of the good life. Indeed, most lives are lived according to a mix directly relatable to these central capacities of the inner sensibilities as described in this work. Similarly prescriptions of classic forms of the good life in western philosophy such as the hedonist, contemplative, saint, hero, activist, or stoic can be easily seen as grounded in these inner sensibilities, the human condition, and radiality. Classic Greek philosophy has seen a high-level of the good life as consisting of an overall individually-appropriate balanced fulfillment of the central qualities of the inner sensibilities in: well-being and happiness, beauty, truth, goodness, justice, autonomous freedom and free functioning, and the flourishing of these in our self, others, and our earthly environments.

More fundamental views of the good life need go to the fuller ground structure and range of the inner sensibilities and human condition than do these identifications. The rough outlines of human nature and the human condition given in chapter thirteen, for example, indicate central qualities and constraints of human existence within which our good life must operate. These include: biological primals, intrinsic pleasures of health, vitality, fun, and humor, the needs to grow and adapt; intrinsic biological, interpersonal, and spiritual needs; our primal necessity to act (which manifests in economic terms in modern society), our functional nature, ability to create; our radial needs, open-ended adaptation and exploratory self-corrective growth, radial transpersonal universal and transcendent involvements, pleasure, friendships, and love; purpose, place, and quests; learning and internal growth of self; highly varied realms of transpersonal achievement or service; deep personal integrations in our internal reflective space.

Maslow sees growth-motivation as a central driving force in the good life[6]. Any person's attempt at a good life plays out according to one's history, abilities, and circumstances.

Examples of good lives or attempts at good lives, mostly organically grown and not artificially structured as are these concepts, are present all around us in anyone we see, and in our selves. The diversity and potential for personally fulfilling good lives with acceptable moralities are virtually unlimited.

The self-other question is central to any level of commitment of morality, and is also reflected at any stage by the relation of one's good life with one's morality. These latter two should be mutually reinforcing. A naturally radial self-centered good life needs to include some individually appropriately altruistic moral attributes, both for simple social survival, and for intrinsic-inherent inner integrations. Morality itself emerges in the practical needs of life and the inner valuative integrations in themselves. Symmetrically, because of one's special closeness to one's own total experience of goodness by intrinsic radiality and as central to all his functioning, one's own good life should be a major consideration in any wider morality, even if one eventually gives this good life over for universal or transcendent concerns. Beyond these morality has its own deep transpersonal rewards and wholeness. Those who cultivate goodness in itself find the fulfillment of self in its transpersonal universals and transcendencies.

Generally, in typical moderately good lives, the good life and morality should mutually support each other in a balance particularly established within each given individual. Perhaps the best way to show this is to embed the good life within a generic componential description of morality with the understanding that the good life is a primary desideratum in itself and recurrently selects its morality as well. Such a generic universal morality could consist of: cultivate good inner attitudes, cultivate the personally fulfilling good life in oneself, balance self and others across primal radiality, ameliorate/counter badness and evil, cultivate and support good valenced thoughts and good works. These components reflect the map of fountal streams shown in Figure 14: the first recognizes the directive power of intrinsic inner reflective space, the next two reflect the centrality of our radiality to anything we do and especially to morality, and the last two reflect our open-ended placement vis-a-vis the intrinsic sides of goodness itself--the good and the bad.

Representative good valenced thoughts can include: universal communal enlightenments (understanding, education, beauty, arts, justice, law, goodness, churches); community goods (internal and external security, community functional structures, order, communal well-being, vision, progress); world peace, stability, process; universal human rights; and personal, domestic, and local enlightenments. Many universal-transcendent values and qualities of being such as suggested in Table 6b can also be taken as fundamental good valenced thoughts. Good works can include anything that aids the well-being of others or advances the positive effectiveness of governmental or altruistic organizations, or the communal enlightenments (goodness, truth, beauty).

Several central streams of badness common in human communities can be recognized as: injustice, deprivation, disadvantage, underdevelopment, slavery, tyranny, war; provinciality, narrowness of mind, chauvinism, intolerance, vengefulness, dogmatism, judgmentalism, conditioned bias, prejudice, bigotry, gratuitous nastiness; envy, malice, abuse, oppression, violence, anti-goodness, anti-sensibility--evil; excessive self-interest, greed, petty advantage-taking, cheating, dishonesty, falsity, vanity, anger; ego-attachment, self-aggrandizement; misdirected sensuality, lust, vice; aggression, crime.

Personal balancing of one's primal placement in self-interested radiality with the well-being and wants of others is a major, even dominating feature of any morality scheme and most of the world's religions, and ultimately, of most of the problems of the world. Primary principles are respecting others, fairness, justice, avoiding abuse and harm, carrying one's own weight, tolerance, making allowance, acceptance, accommodation, giving and accepting redress or apology, and so on are fundamental to morality and necessary foundations to a smoothly functioning community.

Beyond these rudimentary levels, the central existential moral choices individuals face often come down to a question of the degree of personal self-sacrifice of one's life energies, interests, and pursuits of pleasure, comfort, achievements, on the one hand, and fulfillments for others or communal or altruistic interests, on the other. The contingent circumstances, inner self-reliance, opportunities, wherewithal, and place and nature of all participants color and influence such situations.

The heart of anyone's morality in the end refers to one's deepest inner feelings and attitudes as to the good, as one has found or developed these within inner reflection or intimations, and has adopted or revised earlier spiritual training and conditioning, all in conjunction with one's direct life experiences. These may vary from those pathologically disaffected by early abuse or neglect, through sparse minimal incomplete tendencies of partial control or other-directedness, through any degree of systematic wholeness, to the holiness of the world's saints. Their variety is unlimited and in the hands of genetics, life experience, surroundings, circumstance, and one's inner reflective creation.

14.5.1 *Morality*

The key of morality seems simply to hold and show respect, empathy, compassion, and non-harm to others, and to balance self-interest with self-sacrifice for others and universal or transcendent concerns. This in fact, is partially natural, often rewarding, and, in moderation, not difficult in normal circumstances. Coupling this with a vocation or central interests in useful, universal, or transcendent themes makes a valuable and good person on most scales.

Religions, of course, and rightly so, seek deeper, fuller cultivations of the good. There is much essential common ground across the religions of the world, especially regarding the humanistic basic decencies, but also important differences of emphasis and substance. Common ingredients include: fountal experience in the inner reflective space and meditative contemplation, the sense of God as the ultimate goodness and power behind all things, devotion, primal compassion; service of others, community, universals, or transcendents; transcendence of smaller self-centered concerns, amelioration of karmic disturbances, preciousness of life and non-abuse of others. Other qualities are indicated in the tables of chapters three and eleven.

The medieval Christian church identified seven essential virtues as justice, courage, temperance, and prudence (drawing from classical Rome), and faith, hope, and charity (drawing from Jesus). It also identified seven essential sins as envy, anger, pride, lust, incontinence, gluttony, and sloth. Buddha identifies ten sources of human badness as murder, theft, adultery, lying, slander, abuse, idle talk, covetousness, hatred, and error.

Hinduism sees the solution of the problem of karma in the principles of blanket non-attachment to one's smaller self and ego, and non-attached universal-directed action. Jesus sees it in his central teaching: love thy neighbor as oneself, forgive and pray for those who hurt you, return good for evil, turn the other cheek. Buddha sees the answer in the conditions of the human heart and the peace and transcendent understandings of deep meditation. The Chinese see it in the harmonization of one's self with nature and others. The philosophic stoic sees the answer in resigned reflective contemplation.

Saintly lives in many traditions are closely related to: primal compassion, the disturbances of karma and radiality in human affairs, the fountal experiences of inner reflective space, and the subordination/alignment of self with God--seen as the ultimate goodness and power underlying one's being and all else-- and self-sacrificing service of others, especially the needy.

Guilt, shame, and blame are highly potent players for worsening in the areas of the good life, morality, and karma. They should be avoided in favor of understanding and constructive redress.

14.6 Morality, Darwinian Selection, Cultural Evolution, and Species Survival

The variables relevant to the influence of morality in human evolution are encompassed within an all-inclusive ambient environment (say, the earth's surface or any appropriate region of it) and three levels of groupings of all living beings--individual beings, populations, and species. A species is a collection of all those beings who may successfully reproduce together; a population is taken here as a subgroup of a species defined by the inclusion of most mating of its members and the sharing of a gene pool in possible distinction from those not in the population. The term gene pool may be applied sometimes to a species or population. All three levels of groupings incur pressures from the ambient environment which try the coping powers of all their characteristics.

Darwinian natural selection is a description of the mechanics by which certain characteristics relevant to the coping powers of a gene pool in its ambient environment become more or less prominent over generations[7]. Significant changes in the characteristics of a gene pool may occur over successive generations by the interactions of environmental pressures on the gene pool resulting in variation of a characteristic across the gene pool. The main mechanisms are 'selection pressure' and 'differential reproductive advantage'. Selection operates upon a base wherein most genetic characteristics exhibit a certain amount of variation across members of even stable gene pools. Also genes characteristically incur various mutations in reproduction or radiation trauma, some of which significantly affect the coping powers of the individual carrying the mutation. Suppose a given stable gene pool enjoys a certain moderate variability in a given characteristic (mobility, skin color) when a step change occurs in its environment (influx of a predator, weather pattern shift to high incident radiative sunlight). These environmental changes are selection pressures which favor protective characteristics (high mobility, dark skin color--blocking radiation). Those genes which produce these favored characteristics in their offspring enjoy a differential reproductive advantage over the particular parent gene in the gene pool in that those holding the parent gene are less likely to successfully reproduce themselves. Gradually over generations the percentage of the gene pool holding the advantaged characteristic (high mobility, dark skin) increases. In practice, the change can alter the population remarkably quickly. Such genetic changes are sometimes sufficient to disallow reproduction between members of the new gene pool (advantaged characteristic) and members of the old gene pool or members of a different population of the same original species, thereby constituting the emergence of a new species. This process is the Darwinian natural selection of characteristics and evolution of species.

Darwinian selection acts through the genes of individual members and their individual reproductive capability within populations subjected to environmental pressures on coping abilities of the genes. We may take this as the first of three levels of evolutionary change.

Human beings exhibit coping characteristics within their inner sensibilities in both genetically determined and plastically learned forms, in biological primals of life-support and appetitive drives, adaptive integrations, and higher inner representations, reflections, and valuations--both self-serving and transpersonal. We indeed thrive by our partial transcendence of genetic prescription in itself. This shift from genetic prescription to plastic adaptation and then to detached inner reflection is the determinative hallmark of our species. Our evolution as a species builds on the basis of Darwinian natural selection of genes, but in fact is fundamentally different in mechanism from it. Our plastic constructions of understanding and farming, the building of advanced tools, weapons, cities, and industrial processes are not contained in nor transmitted in genes but in handed-down traditions of communities[8]. Moreover, most of these serve the survival of entire populations rather than selected individuals or subsets. The term 'cultural evolution' is appropriate to this realm of plastic constructions. We may take the term culture to apply to a species or population as appropriate.

Over the one hundred and twenty-five thousand or so years of our existence as a species we have survived and come to flourish through our ever-increasing technological control of our environment, and our progressively increasing understandings of this environment and each other. We can say that high abilities in these areas comprise 'differential productive advantages' to a culture (population) as compared to the genetic differential reproductive advantages of Darwinian natural selection. The effects of human productive evolution may appear in broad outline similar to those of genetic reproductive evolution.

Humans thus exhibit these two levels of genetic (reproductive) and plastic (productive) constructions and evolution, and many plastic productions facilitate genetic reproductions. Cultivated values, including morality, now may be considered as a third and yet higher-order realm of human qualities. Morality, rooted in the inner sensibilities of good and bad as mapped for example in the fountal streams of Fig. 14.1, emerges from both genetically prescribed and plastic foundations. Further, morality exhibits a wide range of qualities with both superficial and fundamental limiting regions. The more superficial moralities can be local, provincial, particular, self-serving, divisive, intolerant, self-righteous, or ignorant, thus introducing interpopulation rivalry and hostility. At the operational level, these seem to serve little identifiable positive effects except perhaps to solidify a local group.

The more fundamental levels of morality engage qualities of goodness that are supportive, inclusive, and non-judgmental. They include things like: primal compassion, deep emotional respect of all life, abhorrence of violence and abuse, tolerance, respect for higher sensibilities of truth, understanding, reason, grace, beauty, fairness, justness, many forms of goodness, goodness in itself. These qualities of morality seem universal in the deeper reaches of most religions and spiritual traditions of the world. They also seem rooted in the deeper genetic essences of the higher inner sensibilities.

At the evolutionary level, these fundamental qualities of morality tend to support and protect life, and to increase regard for the well-being of others. In this they should effect some more or less direct positive cultural effect on the survival of life. At perhaps more subtle levels, such attitudes tend to increase harmony and decrease discord within populations. They also help guide significant population and individual decisions to fuller inner reflection. They increase the quality of life, the depth of meaning in life, and bring value more explicitly and fully into life and its decisions. All these kinds of things may be seen as a third level of evolution beyond reproduction and production which we might label harmony. We could say that harmony facilitates production which facilitates reproduction. We can also note a correspondence of these levels of human evolution with the human inner sensibilities themselves: reproduction with the biological primals, productivity with adaptive integrations, and harmony with the inner valuative representations. (It is also interesting to note the the so-called cultural wars of the previous decade can be associated with this third level of evolution.)

The fundamental qualities of morality are related to and serve reason and understanding, but seem rooted more deeply in regions of the dimension of goodness itself. Moreover, they are personal but also and especially transpersonal, universal, even transcendent. They serve an individual, but also others, one's population, and the species as a whole. They tend to reach out inclusively to all human life or religious boundaries. Indeed, they reach out beyond humanity to all life. It is notable that some fundamental moral qualities have universal appeal and therefore may thus have some linkages to genetic factors, and so, inherency.

Note finally that many human qualities, including values of the good life and morality, may make valuable, significant, desirable contributions to the quality of human life even if they wouldn't increase survivability, as indicated in chapter four.

14.7 Closing Comments

In this chapter we have used the foundations, structure, and nature of the inner sensibilities, the full ranges of human value, and the human condition to discuss the ground structure for an inclusive and broadened human view of both the good life and morality universal to all humans. This approach to the good life and morality seeks a sort of 'natural law' rooted in generic intrinsic and inherent features of human existence, but given over to individual particularity, diversity, plasticity, open-endedness, and ultimate individual autonomous optional cultivation, judgment, and direction.

Essential factors here are to identify root generic qualities universally common to the inner sensibilities and human condition that all humans can share, and to leave ultimate judgment on particulars open to autonomous individual seats of judgment. The approach counsels the avoidance of overprescription, overproscription, and generally also of the explicit focus on content of most particular half true valenced thoughts.

Western philosophers to the present have held that the function of reason is to guide living; we add that this function is guided by the fuller extrarational sensibilities including especially value and the aesthetic, and must, moreover, continually give itself over to reevaluation in current existential circumstances. All externally recorded rational templates are products of a rationality fed by a fuller sensibility, and are subject to such continuous reevaluation in complex mult-dimensioned living instances. All are simply fodder for here-now judgments and choices in such ongoing living.

According to the perspectives of this view of our common foundations, most established religious views reflect significant trans-secular spiritual truths, but may be sometimes overly unbalanced, particularized, or overly prescribing or proscribing. According to these same perspectives, many or most secular views are too small: overreduction, fractionalization, false conundra, overstatement, lack of perspective and balance, narrow self-centeredness are not uncommon in the communities, nations, media, and cultures of the world, and in some corners of academia, religions and churches. There are needs for wider understandings, mutual accommodations, and integrations across individuals and all these larger-order human groupings.

This chapter has offered a cohesive universal view of the personally fulfilling good life and of morality within a ground structure of the human sensibilities, condition, and potential, and the fountal streams of good and evil. This is offered as a grounding perspective of human inclinations in these directions, and a sense of the homologous commonality of us all before and within these questions. It thus allows a universal common ground across individuals, ethnic and interest groups, nations, and cultures. It also renders the great plastic insights of all the world's spiritual and philosophic teachings accessible to the ultimate judgments of individuals in their own particular mixes of things.

Endnotes

1. Bowlby1969, 1973; Schore2002; Lehman and Jegtvig 2004
2. Lewis, personal communication2005
3. Noback et al1991, Sarnat and Netsky1974, Yew et al2001
4. Campbell 1959-67
5. Edwards1967
6. Maslow 1968
7. Gould2001, Tudge 2000
8. De Wall2001

Chapter 15

Contextual Energization Fields

This chapter applies the considerations of the previous pages to the natural everyday living of individual human lives. The goal is to identify the central inner and outer factors which drive and constrain our living of our lives. These determining influences are labeled the contextual energization fields. Contextual here indicates those qualities and circumstances that surround, underlie, and influence human life. Energization underscores their ability to force and influence human life and bring about change. These contextual energization fields can be internal, external, or mixed. They are described here in universal generic terms applicable to any human in any place or time, yet purport to identify the nature of those central regions where our richness of individuality flowers and our key individualizing decisions are made.

This project requires a higher-level operational perspective than considered thus far-- more richly developed personal descriptions, and more visible existential dimensions. The basic fabric of the way of placing things is highly useful here, but its constructions need to be subordinated to this different perspective and its needs.

This project is useful because it explicitly recognizes and places the various internal and external influences of our surroundings which collectively drive the unfolding of our and others' lives and circumstances. In this it helps us better understand and contribute to these continuing unfoldings. It informs our attitudes, expectations, and intentions, and gives ground structure to our engagements or actions.

This chapter will first collect and identify the various elemental contextual energization fields from the earlier discussions of this book, then briefly characterize each of these. These will then be organized and discussed in terms of an operational summary figure of internal and external composite contextual energization fields of personal human living.

15.1 Elemental Contextual Energization Fields

Eleven primary contextual energization fields are shown in Table 15.1 in three groups. The first four energization fields are first-level realms of the way of placing things: the whole of nature, the worlds of man, communal enlightenments, and the sociopersonal. These are seen here from a human operational perspective, and this is underscored by the explicit recognition of additional subrealms of natural and world energy systems added to the whole of nature and words of man, respectively. The inner sensibilities are described in terms of personal living, inner depths, and vision and direction all of which relate more directly to our inner operational control (all seen here in a fuller existential sense from within). More of its qualities are included implicitly within human nature which is one of a third group which consists of four elemental fields central to existential life: human nature, the human condition, primary realms of experience, and fountal streams of good and evil. All of these have been introduced in the two previous chapters. Later the first and third groups of four each and personal living will be taken as external contextual fields because they necessarily involve considerable external engagement in almost all instances. The inner depths and visions and directions constitute the two solely internal contextual fields.

15.1.1 *The individual contextual energization fields*

Table 15.1 indicates the nature of each of these individual contextual energization fields by several subheadings.

The whole of nature and its natural energy systems comprise the physical givens of life. They are the ultimate source of all our material needs from the very breath and matter of our living bodies to our nutritional and ambient needs. Our genes are molded by its textures. Our lives warmed or tossed asunder by its slightest physical vagaries. We are revitalized by retreats to its kindred wild spaces, though most of

Table 15.1 Contextual energization fields of human living.

Communal enlightenments
-oversight
-aesthetics. arts-letters.
-understanding. reason. truth
-value. goodness. religion

Personal living
-interests. leisure. play. Outings
-business. mechanics of living
-circumstances. Intimates
-quest. work. dharma
-inner depths

Visions and direction
-intentions, plans
-quest
-purpose. dharma. Role
-inner reflective space

Inner depths
-harmonizations
-transcendence
-spirit
-mind
-body

Whole of nature
-natural energy systems
-sociobehavioral universe
-mentoexperiential universe
-biological universe
-physical universe
-ultimate mysteries

Worlds of man
-worldly energy systems
-places of the world. travel
-current living world
-history. social theory
-world cultures
-settings. origins; rudiments

Good and evil
-[fount. benevolence. empathy. friendship. love. biology. radiality. malice. bigotry. provincialism. community. enlightenment]

Sociopersonal
-arenas & intersections
-social conventions. Protocol
-karma
-people
-individual life

Primary realms of experience
-death
-time. change. aging
-pain. loss
-motivational complex
-feelings. emotions. Instinct
-comfort. pleasure. joy. love

Human nature
-multiple inner causality- [biological. adaptive. imaginative]
-multiple operating levels- [self. group. culture. universal. transcendent]
-mix of mode types- [hedonist. engineer. artist. scholar. moralist. politicain]
-partially original- [genetic. learner. creative]
-variable inner: size, depth, breadth

Universal human condition
-interpersonal bonds, communities
-partially free conscious quest
-plastic open-ended imagination
-incomplete, mixed, radial
-rooted in life & want
-necessity to act
-intrinsic viewing point
-obscurity of origin, nature, role

us in modern civilizations make little use of this. Collectively, we must almost always simply adapt to nature's nature, but we sometimes protect ourselves against it at a price. As we are becoming more numerous and technologically advanced we impinge increasingly invasively on this nurturing ground stuff of our existence. It is our most primordial contextual energization field with many largely unnoticed manifestations. Nature is also the focus of most and ultimate grounding of all science.

The universal human condition is also a baseline of human existence. Its qualities, constraints, and influences are ever-present but usually quiet and often unnoticed in the background of human affairs. This energization field wields largely constant universal, neutral, and mostly impersonal constraining influences and limited freedoms on oneself and all humans. Mostly we all stand comparably under its fallout, but significant deviations of vantage are incurred by many. Again, we are usually called to adapt to its nature as best we can.

Human nature and the primary realms of experience are slightly but only slightly more malleable staple ingredients of human existence. These thrust themselves into our lives often at or beyond the limits of one's or anyone else's control. They often manifest more personally, are more invasive and value-charged, and more deeply experienced than most other contextual fields. They are close to the warp and woof of human existential life, are the inescapable existential fabric of human living. They are the stuff of much poetry and song.

The worlds of man provide both the practical necessities of security, order, economic productivity and distribution, and great enhancements of human living and the human potential. They serve as a pressure cooker for human activity and sociopersonal interacations. Their energizations exert strong large-scale pressures for good or ill on the circumstances, opportunities, freedoms, and fortunes of their constituent human members. They are the stuff of history and current events with all their collective achievements, struggles, debates, wars, oppressions, progess, and enlightenments.

The sociopersonal dimension including its external face of individual life [(Table 12.1(b)] labels the important collective activities of work, play, and interaction which engage most of the lives of most people, and often fill and govern much of our inner mental space as well. This is the home of the work, business, professional, and practical sides of life, its worldly achievements, pleasures, adventures, dangers and karmic involvements. There is tremendous idiosyncratic individuality in this realm, even though outer structures and circumstances may be similar. This generalized sociopersonal realm which is rounded out in the last section is especially the central stuff of much literature.

Good and evil, represented here in a dozen fountal streams, are powerful charges by which people hurt, help, or comfort others or are so done by them - - carelessly or intentionally, cruelly or benevolently. Their badness adds an electricity to human affairs whose voltage can range from minimal to what may seem infinitely high. Its goodness is a warmth and a balm. We all must learn to deal with these as they permeate the give and take of collective existence. They are the ground stuff of ethics, morality, and crime, and major factors of spirituality and religion.

The communal enlightenments are the collected realm of concern and accomplishments in the development of the inner sensibilities especially along the lines of the valuative representations - -generalized: aesthetic perceptions, reason and truth, and value and goodness, and the support of communal well-being and governance. These manifest especially in the arts and letters, scholarship and science, humanities and religions, communal functional structures and leadership. They are the collected realms of human accomplishment and sensibility that come closest to the higher inner sensibilities of individuals. They have strong existential influences and enhance the depth and quality of individual and collective life. They are the spirit and stuff of culture, community, civilization, and individual education and growth.

The dimension of personal living is the array of qualities and activities which constitute the daily living of life by a given individual. Its subheadings shown in Table 15.1 can be adapted to human lives across all space and time. It is the contextual energization field through which one's inner sensibilities engages most of the other contextual fields. It contains all of ones actions, mindsets, inner attitudes and intentions. It is a powerful energization field especially for the individual, but also for others whom one encounters. These are the externally-directed contextual energizations over which we have the most control and which ultimately influence ourselves more than do any of the others. It is the construct which most closely relates to the idea of the self.

Inner depths and vision and direction are two vitally important internal contextual energization fields. It is in these two that our inherent qualities of humanness define themselves for us and from them that our humanness extends through our personal living outward into the communal world. Their central ingredients are indicated in Table 12.1 (b) and Table and Fig. 15.1.

15.2 Fundamental Contextual Energizations of Personal Living

Figure 15.1 shows a broader integrated view of the contextual fields discussed above which better reflects the comparative nature of the fields and an individual's existential place within them. Particularly Figure 15.1 groups the central individual fields of personal living, vision and direction, and inner depths together (composite individual field), and also the kindred fields of the human condition, human nature, and primary realms of experience (composite condition field). The composite individual field is richly idiosyncratic and plastic and stands among a half dozen readily recognizable operationally shared fields with which it engages with variable individuality within generic commonality.

Figure 15.1 provides a fundamental cohesive framework of the undergirdings of the influences governing the overall circumstances and force fields of human life as the basis of a systematic consideration of ideas regarding the overall circumstances of human life and living. For example, the tenets of, say, post-modern thinking (see Chapter 16), or of wisdom can be meaningfully approached from this perspective. It oes this in terms which are identified and discussed more fully in preceding chapters and Chapter 16.

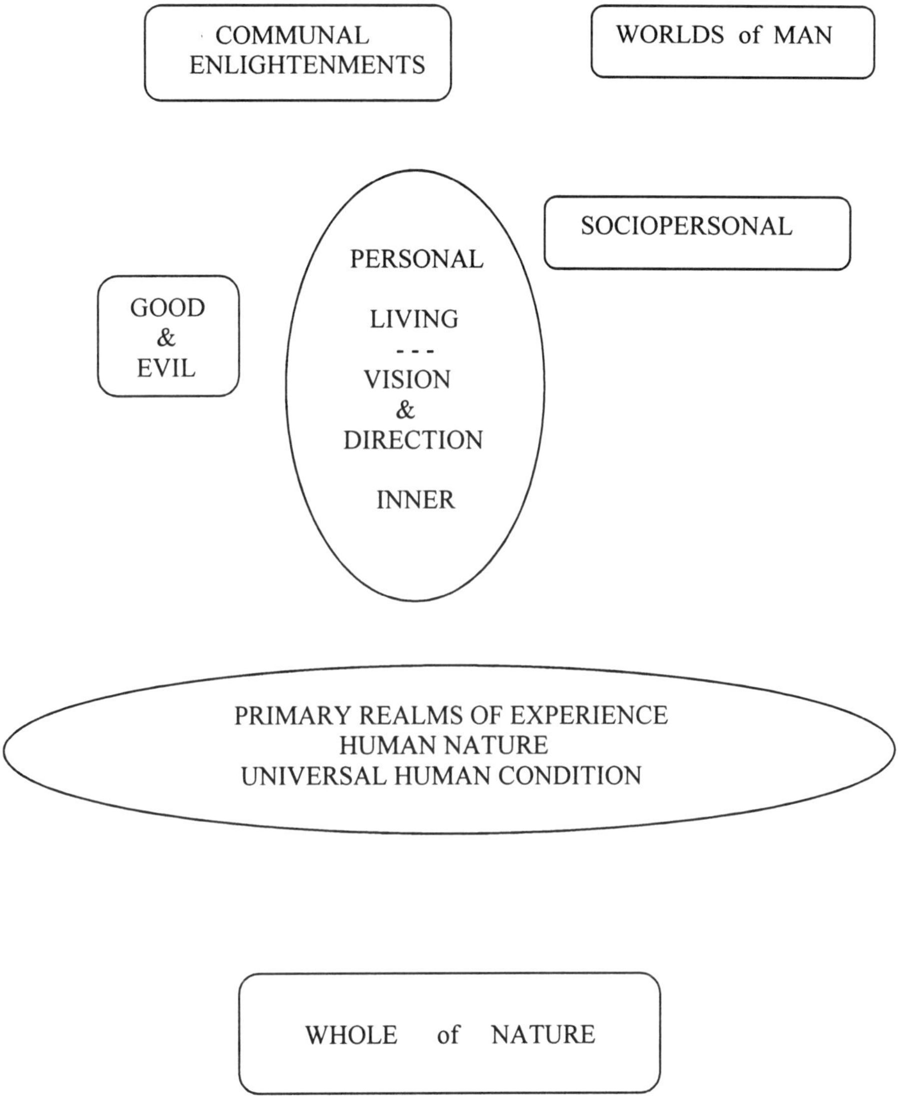

Fig. 15.1 Contextual energization fields of personal human living.

Chapter 16

Implications and Interpretations

This chapter briefly outlines interpretations and implications of the inner sensibilities regarding: scholarly research and general interests; academic philosophy; the human potential; fuller ambient contexts; and spirit, metaphysics, science, and religion.

16.1 Particular Implications of This Work

16.1.1 *Knowing, neurobiology, and epistemology*

These topics have been broadly treated in chapters two, eight, nine, and ten. They have significant implications in neurobiology, epistemology, and the philosophy of mind.

16.1.1.1 *Neuroscience*

The view of knowing presented here includes an overall outline of the essential operational neuroelectric engineering of the brain regarding the cognitive operations of truth and knowing. With these pictures of coordinated neuroelectric patterns of cognitive representations and the idea of their match-mismatch tests one may construct large-scale computer simulation studies to explore qualities of cognition and knowing in brain-like model studies[1] for comparison with increasingly sophisticated non-invasive experimentation in humans. With the help of such models, one may explore higher level psychological principles of cognition such as the theory of 'cognitive dissonance', as well-formulated by Festinger[2], in relation to overall cognitive brain dynamics. This theory is highly akin to the powerful minimization theories of classical physics, and may well be a prototype for similarly high-level minimization processes throughout the rich variety of the brain's neurobiological control systems.

16.1.1.2 *Epistemology*

This work's view of human knowing, through its placing of rational knowing within wider extrarational cognitions and other sensibilities and its explicit typology, supports Haack's key concept of inherentism in contemporary epistemology[3]. Haack's concepts significantly expand our sense of human knowing by relaxing some unduly restrictive features of a narrowly-conceived rationality. The grounding of knowing in neurobiology given here offers additional dimensions to epistemologists, as for example in the relations of archetypes and other intrinsic neural interconnection patterns to knowing, perhaps as examples of synthetic a priori knowledge. This view of the neurobiological grounding of knowing sees plasticity as the robust quintessential component of the human organizational scheme of operational living--plastic learning replacing genetic prescription. It would seem that this open-ended plasticity should be given higher consideration in epistemology than it receives.

16.1.2 *Foundations and function of consciousness*

These problems are introduced in chapters two and five, and treated more fully in Chapters 6, 7 and 9. They have significant implications for neurobiology, theoretical physics, the philosophy of mind, philosophy, and anyone interested in the nature of man.

16.1.2.1 *Theoretical physics*

The contradiction between the apparent partial autonomy of

consciousness and a brain whose neurobiology does it all by classical physical law is an enigma at the heart of consciousness and its role in the brain. This work argues that this enigma is our most significant indicator as to the fundamental nature of consciousness. The enigma suggests that at least something of consciousness may be somehow beyond classical physics or beyond the brain, and that its partial autonomy is fundamentally associated with energization--the ability to exert a force, to cause motion or change. It would seem highly appropriate for theoretical physicists to take a share in this enigma. Surely there should be some place for it within physics' grand quest for a 'theory of everything'[4].

16.1.2.2 *Brain organization*

The theory of consciousness developed in chapter seven also has implications for overall control in brain organization, including a possible minimization principle. Its detailed localization in brain tissue is less specific because of the uncertainty of the nature of consciousness. One important finding is its possible relation to metabolism through the medium of glutamate control of glial cells as described in chapter seven. Again the theory of chapters six and seven can be developed to help guide and interpret brain imaging experimentation on localization, actions, and associations of consciousness. It seems important that experimental and theoretical exploration be undertaken in quest of the relation of conscious autonomy to the energization of neurobiological activity. This is surely a very difficult and highly obscure challenge, but may well entail a watershed convolution in our currently overreductive collective understanding of the functional organization of the brain and of human nature itself.

16.1.3 *The fuller inner sensibilities and its qualities*

These topics are considered in chapters two, three, four, five, and nine. They have significant implications for philosophers, artists, humanists, religious scholars and practitioners, and any others interested in the nature of man. The implications of the inner sensibilities for academic philosophy are considered in the next main section below. The inner sensibilities as antidote for the overreduction of the human potential is considered in the third section.

Artists and the aesthetically-inclined[5] may find interest in our view of human aesthetic perception as a primitive type of intuitive cognition in quest of a sense of the essential. Artists, humanists, lawyers, and others interested in human nature may find interest in the grounding of aesthetics and beauty, understanding and truth, value and goodness, and justice in neurobiology.

16.1.4 *Fuller ambient contexts*

These topics are considered in all chapters of Part C. They have significant implications for the guidance of individual and collective living, and the problems thereof. They are considered more fully in the fourth main section of this chapter.

16.1.5 *Value, good and evil, morality*

These topics are considered in Chapters 2, 3, 11 and 14. They have significant implications for neuroscience, the philosophy of mind, ethics, religion, and the common interests of individual and collective life[6]. They are considered more broadly in the last main section of this chapter.

16.2 The Inner Sensibilities and Academic Philosophy

Philosophers looking past the conundra of post-modern thinking[7] to broader interpretations of the nature of man may find interest in the expanded integrative view of the properties of the brain presented here. Our view of first principles is indebted[8] to Descartes. Our overall integrative approach is reminiscent of eighteenth-century ideology (for example, Kant) updated to interface with contemporary neurobiology. Our view also reaches back to earlier post-Darwinian approaches in organic biological-like systemics (for example, Bergson's *l'energie spirituelle* and *elan vital*). Our approach is most kindred with that of Freud. Freud's triad of id (biological primals), ego (adaptive integrations), and superego (social repressions) are more narrow and competing capacities, as contrasted to our more broadly conceived inner sensibilities with their inner reflective space which are interpreted as intracooperative along the lines of Hughling Jackson's evolutionary levels of construction. Nonetheless, our views differ from all of these in many particulars.

16.2.1 *Twentieth-century philosophy*

Akin with movements in poetry, literature, and the arts which broke with traditions in late and *fin de siecle* nineteenth-century, and especially for philosophy in the thought of Nietzsche, western philosophy of the twentieth-century began to define itself around a break especially with metaphysical questions which had been a central feature of philosophy since the early greeks[9]. The new views[10] wanted to make philosophy more scientific, to get past metaphysical dualisms (material and spiritual, body and mind), and claimed that philosophy should not study things which could not be clearly demonstrated to exist. Wittgenstein, who studied logic and mathematical structures under Russell, claimed that language was the only foundation of all true statements about the world and that not much in fact could be reliably said. Husserl stressed consciousness rather than language as the alternative foundation to transcendental posits of metaphysics, and claimed that human meaning emerged in its interactions with the external 'life-world' in daily living and manifested in intentions and actions. Ayer advocated the restriction of philosophy to issues which can be verified by empirical observation, effectively a reduction of the field to philosophy of science.

These views gained wide support within philosophy, defining a movement known as logical positivism. This view has permeated the philosophy of mind[11], science, and neuroscience throughout the century.

Logical positivism can be considered the ultra-conservative right in terms of which all later work defined itself (either supportive or adversary).

Several significant forms of existentialism stressed the vital pre-eminence of conscious experiential life with its feelings, intuitions, drives, contradictions, and uncertainties, identifying itself as the far left of philosophical thought and associating with literary expression as much as with philosophical treatises[12]. Existentialist thinking centered within the whole person, a philosophical concept Ayer advocated from his different perspective.

Many, perhaps most, other philosophers undertook the critical analysis of human cognitive capacities and other inner sensibilities throughout the century, defining by the beginning of its last quarter a dominant view of 'post-modern thought'[13]. This wide range of criticism has defined: a nihilism wherein, following Nietzsche there is no reliable basis for value, goodness, or truth--these are seen as sociopolitical agreements by individuals whose thinking is constrained by sociocultural conditioning, personal interest, and unavoidable bias of perspective--and 'god is dead'; all judgments are necessarily relative to all these factors, there is no objectivity, and there is no basis from which to adjudicate among views, so all views are equally valid; our processing of knowing is pervasively flawed with imperfections, so the best we can do is a vulgar pragmatism; the violence of human life and history and vile currents in the human heart render pessimism the only sensitive attitude; the deconstruction of viewpoints and texts to reveal the nature of their biased perspectives and ulterior purposes is the only valid task of philosophy. All this is seen to take originality and creativity away from the artist or any individual and replace it rather with their sociocultural place and conditioning. This deconstuctionism was developed by many others who identified interpretation (hermeneutics) as a (the) major dimension of philosophical thought, taking the emphasis on language to the extreme: 'the text is everything'.

Other philosophers have contributed more constructive original concepts regarding the nature of human inner sensibility[14]. The pragmatism of William James defines a robust adaptive plasticity whose knowing which, although contingent and plastic, is strong 'all around and up to the present moment'. Levi-Strauss has developed the idea that the mind constructs categories of things according to the way we do things. Quine has echoed this in categorizing memories as semantic (declarative) or action oriented (procedural), and has also identified a 'web of belief' as a key organizational quality of the operative mind. Gadamer has revived respect for practical reason. Haack has described a level of inherent knowing and a concept of foundherentism. Searle has identified biology as the determinative foundation of the mind. Relatedly, but further to the left are the descriptions of Bergson of the *l'energie spirituelle* and *élan vital* of the conscious mind.

16.2.2 *Critique of the four main movements of twentieth-century philosophy*

Logical positivism can be best seen as a component of epistemology, solidly rooted in the concern and study of knowing in itself. Further, it is restricted to hard analytic rational thought, and rooted (through Wittgenstein's studies with Russell) in the most pure, yet most narrow and rigid kind of knowing—mathematics and logic. It defines itself in the model of science. In this, it would seem to forsake the larger roles of philosophy. It ignores the fuller range and capacities of the inner sensibilities.

Post-modernism can be most charitably seen as a collection of critical analytical approaches which help clarify shortcomings in human apprehensions and knowing. It is however, fatally flawed and without qualities for endurance. It is highly overreductive of the full human inner sensibilities and potential. It consists entirely of overwrought inferences from half-truths. It exaggerates its ideas beyond their range of validity. It takes an entirely non-constructive posture, and unjustifiably parades itself as an end (or collection of ends) in itself. It is grounded in the epistemological criticism of hard rational thought, yet doesn't recognize the place and limits of that foundation. It is unnecessarily and groundlessly pessimistic[15].

Existentialism has been vital to twentieth-century philosophy in providing the needed anti-pole to the rigid hyper-analytic postures of logical positivism. It has stressed the important qualities of feelings and extrarational apprehension. It has brought the realm of the full person into philosophy, stressing this overall individual realm in another strong contradistinction from logical positivism. Existentialism is not complete in itself, but should play a significant role in subsequent fuller integration of the inner sensibilities in relation to philosophical concerns. Some of its views also have been excessively pessimistic, restrictive, or denigrating of the human potential, reflecting mostly the brutal oppressions and wars of the first half of the century.

It is remarkable that all these three fields take philosophy into major realms of academic thought outside of philosophy as such, keeping only a sense of rational analysis as home base in philosophy. Logical positivism has related strongly to science, existentialism and post-modern thought to literature. Existentialism has also found favor among religious and spiritual scholars, many of whom have constructed versions of it consistent with their views.

The constructive thinking of James, Haack, Levi-Strauss, Quine, Gadamer, Searle, and Bergson as most relevant to the needed fuller integration of the inner sensibilities sought by this book.

16.2.3 *The inner sensibilities and academic philosophy*

This work's integrated and structured neurobiologically-grounded view of the human inner sensibilities asks philosophers to consider the implications for philosophy of our rapidly increasing knowledge of brain organization. Simply the concept and overall qualities of linkage to neurobiological processes can introduce significant new dimensions to philosophy generally; more particular concepts can provide fodder for specialty areas. A specific example can help goad and enrich more generalized thinking. A good example of the potential fruitfulness of this for philosophy is the neural network model of cognition developed in part B and its relation to the epistemological views of inherent knowing described by Haack.

This work has offered structured groundings in neurobiology for human cognition, value and the good and bad, and generalized aesthetic perception. These groundings can provide suggestive concepts for expansions or new areas in the fields of epistemology, ethics, and aesthetics.

The view of the inner sensibilities developed in this book also appeals to philosophers to give more credit to the higher sensibilities than it has since these have been pushed to the side along with the rejection of metaphysics. Neurobiological-grounding of the higher sensibilities might help bolster rejuvenation of philosophical concern with these topics. The relation of consciousness and autonomy to theoretical physics suggested in Chapter 6 offers new middle ground for the brain-mind question and perhaps a new approach to metaphysical questions of metaphysics and related topics.

More broadly, many non-philosophers would like to see philosophy reclaim its roles of unlimited generalized rational thought regarding foundational questions of concern and importance regarding human existence and the whole of things. It is such a shame to have our supposedly highest thought reducing itself to handmaidenships of reductive science and to literary criticism. In our view scientific knowledge serves philosophy rather than vice versa as in logical positivism. Certainly hard analytic rational thought should play its role and take its place, but its limitations should not curtail the natural scope of our best and highest collective scholarly thought. Relatedly, might it be useful to have fields: philosophy of human sensibilities, philosophy of the whole person, and sociopersonal philosophy?

A real intellectual shortcoming here is that the voiding out of certain areas on the grounds of uncertainty within a given field (for example, truth, goodness, spirit, God) leaves an incomplete field wherein the remaining areas then emerge as the only players in the field, which in fact may not correspond with the 'reality' presumably under consideration by the field. That is, obscure or uncertain possibilities are dismissed prematurely as non-entities. This is in part what has happened in the philosophy of the twentieth-century. Overreduction is the label we can apply to it.

Major examples of this overreduction are the shutting out of the higher sensibilities by post-modernism, and the shutting out of human conscious autonomy by much of the philosophy of mind and neuroscience. Certainly our higher critical thought as well as our science should recognize the minimal necessary assumption principle (Ohkam's razor) in critical analyses, working hypotheses, and models to clarify limits, boundaries, and uncertainties, yet one needs to remain open to the obscurities of the larger realities beyond the reductions of thinking and models. Twentieth-century physicists, for example, did not imagine that gravity should be disclaimed or somehow undermined because the standard model of quantum mechanics covered so very much ground without it. Similarly, this book holds and has demonstrated, that our higher sensibilities and conscious autonomy are tangible, pervasive energizing qualities of human experience, behavior, productivity, and history. The pre-emptive denial of these without compelling justification seems to me as silly as would have been a denial of gravity by twentieth-century physicists.

16.3 The Inner Sensibilities and The Human Potential

16.3.1 *Division and overreduction*

The lack of recognition of the integrated wholeness of the inner sensibilities and their vital place in human life has been accompanied by a wide and pervasive array of both overreductionism and the fractionalizing absence of bridging recognitions throughout the western scholarly thought of the twentieth century and particularly flowering into the American mainstream in the last thirty years or so. The centuries-old war between science and religion is an example of this[16]. Another is the cultural partitioning recognized in mid-twentieth century by C.P. Snow between the followers of the hard sciences and philosophy on the one hand, and those of arts and humanities on the other, with common culture groping between the two, and religion increasingly marginalized. The overwrought half-truths of post-modern thinking make up a characteristic and pervasively disruptive manifestation of overreduction with many pernicious undermining tendencies.

16.3.2 The inner sensibilities and common ground

This work provides value-neutral ground structure from which large issues of man, spirit, morality, individual and collective living, and the times can be fruitfully approached in terms of their essential underlying energies, constraints, and human significance. In this it provides a common ground structure for viewing many significant considerations in common life, the arts, humanities, and religion, as well as for scholarly fields.

Perhaps the broadest implication regarding cultural integration is that our common inner sensibilities, with their ranges of biological primals, adaptive integrations, and higher sensibilities, is a centerstone, a hub around which all qualities and institutions of human existence revolve, and through which they intercommunicate and indeed share common ground. It is the medium through which we all may come to understand each other and may come to terms with our differences--individually, institutionally, culturally, and globally. It is important to see the entire potpourri of human interests collectively on this widely common plateau. It offers understanding, expanded horizons, common ground, and potential cross-fertilization in many human interests, both in and across individuals, and in and across ethnic groups and cultures.

This integrated view of our common inner sensibilities heightens recognition of hindering divisions in our communities and institutions and provides the inherent basis for bridging these divisions and overreductions. The most common hindering division is that between rational understanding and thought, and the other sensibilities, such as aesthetic feelings, intuitions, or beliefs. The divisions between science and religion, and between the softer realms of arts, humanities, and literature and the harder fields of science and philosophy, for example, are detrimental to holistic personal integration of citizens and the young, and to the overall progression of civilization. The overreductions of post-modern thought are particularly pernicious because in presumably embodying our deepest scholarly philosophy regarding the nature of man, their ramifications pervade widely throughout scholarly thought, and from thence into the common culture and our collective consensual view of what we are and what we might be.

The war between science and religion, for example, is simply the lack of recognition on both sides as to the place of reason within the fuller range inner sensibilities. The religious have a solid base for the essentials of religious faith in the intrinsic human capacity of value, without a need to cling to untenable cognitive structures; perceptive scientists see the larger reality of value underlying the religious spirit beyond the cognitive errors of some believers. Recently Ashbrook and Albright have pointed to common ground beween religion and science on the basis of the 'humanizing brain'[6]. This work has outlined central ingredients of this common ground in chapters three, eleven, and fourteen and will discuss their relations to neurobiological foundations in the last section of this chapter.

Snow's partitioning of scientists and philosophers from people of the humanities reflects the separateness of the analytic and the extra-rational sensibilities. Relatedly, we can identify an essential false assumption in the 'Faustian Conundrum' which despairs at the inability of reason in itself to find its way to value. This reflects simply a shopping in the wrong market. Value is to be found in its own realm. Other significant divisions involve specializations according to the various components of the inner sensibilities to the ultimate weakening of the others within a person--scientists, artists, rationalists, emotionalists, saints, adventurers, politicians, lawyers, athletes, pragmatists, dreamers, hedonists, or the individualities of any of us. There is both rich individual diversity and ultimate underlying commonality of human being. Any of us is a small partial fulfillment of the full human potential, an ultimate minuteness within the grand fabric of life. All these hindering divisions and many similar others are simply those of the blind monks who divided themselves as to whether the elephant was to be taken as his tail, or his ear, or so on. Ultimately, one hopes that recognition of the underlying common richness of our broad common humanity and wide powerful open-ended potential may work towards the amelioration of prejudiced division and alienation.

16.4 The Ambient Contexts of Things Human

16.4.1 *The direction of human living*

Table and Fig. 15.1 provide comprehensive maps of the inner and outer contextual energizations of human individual and collective living as may be readily seen by any individual. These maps are especially useful because by identifying the ground structure within which individual and collective living take place they thereby provide the basis for the intelligent guidance of both. Usually we move through the ongoing changing circumstances of our lives, using more or less automatic conscious navigation according to our learned inner constructions among the influences shown by these maps. The maps can help us better recognize and place the various influences to which our lives are generally subject. Individual instances can be difficult, uncertain, and subject to chance, time, and risk. The maps of chapter fifteen are a useful operational tool for dealing with such instances in terms of the universal energization streams which underlie and drive them. In this our highest strengths are our quintessential species characteristics of individual creative plasticity and valuative representations, and their collective manifestations as our communal enlightenments.

16.4.2 *A positive view of man*

The violence and seeming insanity of much of twentieth-century European history and the deconstructive pessimism of its child, post-modernism, have left a pallor throughout current views of man's nature and potential. Man is widely seen as doggedly fixed in personal uncertainty, perversity, self-interest, and want, and prisoner to the conditionings of his early years and place. All apparent progress in our arts, letters, sciences, and technology is seen as superficial and not touching this inner character whose violence and commanding self-interest can be taken as constant from the earliest historical clues and records to today's headlines.

The constructions of this book lead one to see this view as merely the most sweeping of the many overwrought inferences from half-truths which plague all of post-modern thinking. To the contrary one is led by the present work to see our species as robustly adaptive, well-fitted to our natural earthly environment, and grandly free to conceptualize, imagine, create, construct, fulfill, anticipate, counter, circumvent, modify, adapt, correct, and self-correct regarding most of the conditions, exigencies, errors, and shortcomings of life as we find, embody, and help construct it on earth. The western world has developed such understandings and technology in the last centuries as to render humankind the clear primary determiner of its own future and fate, barring only extraordinary large-scale cosmic or biological catastrophe.

This work leads us to see ourselves in its terms as inherently robustly fitted to biological life on earth through the very construction and properties of our inner sensibilities. We are especially adept at dealing with the vagaries of earthly existence with our largely plastic capacities of adaptive constructions. We are capable of transcending this attachment through our freely plastic imaginative inner reflective space and higher valuative representations. We are not seriously at threat as to natural causes. Our biology is adapted to the physical nature and cycles of earth and its atmosphere. We can cope with most of the normal vagaries of the physical climate on earth, and have come to understand the workings of the physical universe in remarkable depth. We can respond with our collective technology to many large-scale problems, and remain very small only before possible large-scale cosmic events, such as impacts or close passes of a large comet or the eventual explosion of the sun. Within earthly biology, we can generally cope, adapt, learn, and relocate. Our composite abilities, and especially our inner reflections, render us dominant to other life forms. We may become victims of an intractable microbe or some other emergent biological entity, but our biological science and technology is increasingly effective even for such eventualities. Virtually all our threatening troubles arise from short-sightedness, incompetence and human error, disruptive violence within ourselves, between individuals, and within and between communities, nations, or civilizations.

16.4.3 *The central human problems of our times*

All main threats to our existence are directly related to our own species and our own inner nature. Our challenge is to know, temper, harmonize, and fulfill ourselves as individuals, communities, and civilizations, inclusively and without hatred, abuse, and violence. This work suggests these should work out because of intrinsic and inherent tendencies of our inner sensibilities and condition, current and advancing understanding, increasing communication, and current and increasing means.

More specifically, we can identify four main classes of internal human problems: problems within and between individuals; problems within communities; and problems within and between nations. We are threatened by overpopulation as Malthus recognized some two centuries ago. These problems are intimately associated with economic productivity and equity. In our times, they are especially related to newly increased globalization, changing ethnic balances, and marked disparities in economic strength and economic well-being.

This work supports the view that problems within individuals and between individuals can be significantly ameliorated because our own individual inner sensibilities inherently reach for personal integration and the eradication of interpersonal adversity, and our larger personal interests are most fully served by these. It supports the view that problems of race, ethnicity, class, and interest within communities, and problems of rivalry and war between nations or civilizations can be significantly defused because of these factors and because of the great widespread practical advantages associated with solving them, and because the institutional and technological means are available to work effectively on them. A number of factors encourage rivalries and hostilities between individuals, groups, and nations. Perhaps some of these are intrinsic as well as inherent. Not all results of rivalry and competitive struggle are bad. Nonetheless, clear limits can be recognized, and our larger personal and collective interests and our deeper inner sensibilities are better served by keeping such disruptive forces in bounds. It is essential that all should have access to the quest for a fulfilling life, and that the overall health and responsiveness of the society as a whole be maintained.

All this is to say simply there are many intrinsic, inherent, adaptive, and higher valuative factors in the inner sensibilities which encourage cooperation and benevolence among people and collective groups. There are also cultural, inherent, and intrinsic factors which foster adversity, hostility, malice, abuse, and gratuitous cruelty. To this one can say: the good life is both its own responsibility and its own reward; and the best all around and ultimately most deeply effective posture is: for individuals, the recognition and support of the good and the recognition and amelioration of the bad; for governments, prevention, protection, effective enhancing and ameliorating action, and reliably equitable justice.

All our hope for individual progress and fulfillment is rooted in the the deeper and higher reaches of our inner sensibilities (adaptive productivity, understanding, reason, value, justice, empathy), and their collective external manifestations as our communal enlightenments (effective communal functional systems, knowledge, reason, arts, letters, value, religion, spiritual concerns). All our hope for collective progress and fulfillment is rooted in these and most especially, in institutions and leadership with broad inclusive vision and intelligence.

There is a large current need for enlightened visionary understanding and leadership regarding the global interconnectedness, relatedness, and ultimate commonality of all the people, cultures, and nations of the world. These things should be targeted: world peace, security, and order; economic inclusion and participation of all nations within the world; democratic liberties and participation (economic and cultural) for all individuals in all cultures; continual raising of cultural standards and sensibilities of all individuals throughout all underdeveloped and developed nations. The current lack of progress on particularly the essential first of these is acutely tragic because of the rare and likely short-lived present window of opportunity for a global breakthrough which is not being fulfilled. With vision, intelligence, and a steady hand, this or the next generation of western leadership could be the one that makes war a thing of the past.

There is also a need to recognize more clearly the dependence of modern economic society on the full inner sensibilities of its participating citizens. The biggest threats to these goals are :

economic underdevelopment, oppression, ignorance, and provinciality. Provinciality is a particularly pernicious major but insufficiently recognized ingredient especially within developed countries of the west.

Ultimately this window of opportunity in the vast economic strength of the western world can be used to approach a cooperative network of civilizations (a fuller organizational level than nation or even culture) throughout the world enjoying: internal and collective peace, security, and order; a robust basis of functional structures (food-housing-health, economy-commerce-trade, government-law-intelligence-military, order-security-law enforcement & crime, skills-trades-professions, education-enlightenments, entertainments-sports-public events, media and issue redressment); rampant individual freedoms and opportunities for jobs, entertainments, and continued personal and community development; and highly developed communal enlightenments to support and enliven the living of life (communal well-being, participation, service, oversight; arts, letters, humanities, aesthetic sensibilities, music, painting, writing; learning, knowledge, reason, understanding, truth, scholarship; value, spiritual development, good and evil, diversity, tolerance, justice, compassion, meetness).

16.5 Spirit, Metaphysics, Science, Religion, and the Inner Sensibilities

16.5.1 *Monism, determinism and free-will, spirit, religion, and science*

The concept of spirit as an overall temper of conscious existence which engages such qualities as enthusiasm, *joie-de-vivre*, and *elan vital*, is pervasively and universally used in human parlance and reflects an easily and universally recognizable dimension of individual conscious experience. Similarly, the use of the word spirit and spiritual are widely used in another sense as relating to a different range of experiences such as the senses of transcendent existence and meaning, divine intimations, and the like associated with essential mystical experiences discussed in Chapters 3, 11, and 14. This latter use of the word and these experiences are often also taken by many in conjunction with various religious interpretations according to separately received religious teachings or constructions.

In science, this work has follow Ockham's principle of the minimal necessary assumption. In this case, it is the working hypothesis of monism of brain and consciousness. The functional monism described in this work finds no difficulty with any such quality of spirit in either of the two senses described above or any others as verifiable personal experience of any given individual. Interpretation of these experiences as to the nature of total external reality, however, is a separate question. Most religions of the world have taken our conscious experience as part of a larger spirit world which exists in tandem with the observable physical world. Throughout history to the present there have been profound, life-changing mystical experiences in the lives of many, and major religions have captured the faith of many. There has been also much excessive credulity and ignorant overextension of spiritual claims and superstition which continue well into our times. Substantive monists flatly deny the existence of any spirits, divine or otherwise, as existing beyond the mentoexperience of the brain, or surviving beyond the death of the body and brain. The functional monism developed here does not commit, but is open to the array of possibilities listed in chapter six. In all this the enigmatic relation of consciousness to the brain implied by partial autonomy and buttressed by the lack of need for consciousness in a brain that does it all (whose underlying logic is recognized by Ockham's razor) leaves the essential nature and extent of consciousness obscure as yet to science and to rational thought. Outside of rational thought and science there are additional sensibilities, modes of apprehension, and judgments, and different kinds of meaningful experience and truth in the inner experiences, apprehensions, and judgments of any individual. This matter of spirit may remain in these extrarational sensibilities forever beyond our rational ken. If so, this work contains a logic wherein it might be so.

Religion is significantly impacted by considerations of the nature and relations of conscious experience to neurobiology and the physical such as considered in chapters six and seven. First, dualism is almost universal to at least traditional versions of all the religions of the world. Indeed, most religions are constructed explicitly for spirits and spiritual concerns--human and god-like, divine and otherwise. The mainstream religions of the west and mideast (judaism, christianity, islam) see an individual human as existing primarily as a spirit in temporary relation to his body and in potential relationship to one true God, who is omniscient, grand, and good. Human spirits survive the death of their bodies and live eternally in some relationship to God. This last point at least is clearly inconsistent with a substantive monism which sees all experiential and neurobiological activity to cease with the death of the body. It is possible though for even hard monists and others convinced of the close relationship of conscious experience to neurobiology to feel the entire fullness of a religious reverence for life, the good, and God even though accepting a mortality and end of one's total being with all its mentoexperience at death. Any scientific or other truth when fully clear will have to be compatible with the full richness of experience and of nature as we can know it. Any god who is God is compatible with and not threatened by what is true, nor by our incomplete apprehension of truth. Further, the intermediate views between dualism and hard monism indicated in chapter six offer rich and as yet uncharted grounds for metaphysical inquiry.

Separately, the question of partial autonomy also relates to traditional religious views especially of morality. Almost all sects of judaism, christianity, and islam have believed in the centrality of human autonomy and freedom of choice as central factors in the living of a good life. Any scientist or layperson who believes in autonomy may find merit in the greater portions of religious teachings regarding the living of human life while following one's own inclinations regarding the ultimate relationship of the experiential to the physical world, or whether the physical is the full reality, or immortality, or any other questionable obscurity, myth, or dogma.

The denial of even partial autonomy, as held by some strict monists, implies a neural self-sufficiency that seems to reduce individuals to deluded conscious passengers on a driven, buffeted neurochemical machine (his body). Moral codes, for example, a central feature of most religions, seem meaningless without autonomy. Here, a strict monist position of determinism of individual life is similar, though distinct, to the theological moral determinism of Luther and the calvinists who held that God's judgment of salvation or damnation of individual souls was prescribed well ahead of time--predetermined so that an individual's choices of behavior and attitudinal postures had no bearing on this. This theological position seems rather absurd. (But perhaps the calvinist view is not quite so absurd if taken as an empirical reflection that good or bad things may indeed happen to individuals in no apparent relation to their moral qualities. The laws of karma, however, suggest that one's inner attitudes, as well as chance, do indeed strongly influence one's experience.)

Continuing in theological terms, partial autonomy corresponds to energization, which in turn may be linked over to the quality of religious devotion to God as ultimate power, which is a nearly universal feature of religions. Moreover, recall that this question of ultimate energization underlies the two other central obscurities of science identified in chapters one and six as the origin of the universe, and the origin and nature of the self-organizing properties of the DNA and protein molecules of biological life on earth.

16.5.2 *Metaphysics, spirit, science, and religion*

The obscurities, mysteries, and larger grandeur of the whole of things and our deeper inner intimations and questings for a coherent sense of the whole have led mankind since prehistoric times to the senses of metaphysical and spiritual realms of things within or behind the world externally apparent to us. Figure 16.1 illustrates this as a combined metaphysical/spiritual realm or layer to round out explicit inclusion of the obscurities within the three more solidly-grounded realms of physical, biological, and existential (brain-consciousness) reality identified in chapter six as recognized by contemporary science. One may view these four strata, shown in the right side of Fig. 16.1 as objective metaphysics, and the dominion of inner sensibility indicated on the left half of Fig. 16.1 as individual existential (or operative) metaphysics. The two halves of the figure in fact readily interrelate over common ground.

This metaphysical mélange incurs three intermeshed pairings: two faces of metaphysical uncertainty--one in the obscurities, mysteries, and grandeur of total reality (including consciousness and the brain), and one in the feelings and intimations of our global existential conscious awareness; the two distinct dimensions of reflective existential global conscious awareness itself and of rational conceptual knowing; and the recognition of two distinct seats of focus, interest, purpose, and judgment, again that of global conscious awareness in itself and that of rational knowing in itself.

In this *melange a trois*, we can say that knowing can make maps of total reality and of the brain and inner sensibilities including ground structure for the capacities of knowing itself and consciousness. Knowing in itself, however, cannot capture the fullness of conscious experience. This latter is a different dimension, whose shadow but not whose fullness, can be projected within knowing. Global conscious awareness, on the other hand, cannot in itself generate the analytic discriminations of knowing, and resists reducing itself to any overly reductive map of knowing. It feels itself directly, and with a free higher self-loyalty to its own contingencies, preferences, and judgments. It sometimes senses a spiritual nature within oneself, in others, and in intimations of ultimate meanings and God.

In all this, these two dimensions of knowing and of global conscious awareness are sometimes at odds around the surface levels of these metaphysical mysteries. They are also sometimes at odds between different personality types (say scholars or artists or mystics), and sometimes may tend to pull in different directions within any of us. This suggests the speculation of a built-in mutual buffering between these two prominent capacities of our active inner sensibilities, at some limits within their normal overall cooperative relation. It is interesting in this that one may sometimes sense an inner psychological threat if the search for knowing recognition is too much pressed directly onto raw awareness itself. Perhaps relatedly, it sometimes seems that rationality can become overly domineering, tyrannical, perhaps in a sense of unease, rivalry, or envy of the looser free spiritedness natural to global conscious awareness itself.

Senses of the existence of gods or one God have arisen in the minds of humans probably in part through these obscurities, throughout almost all cultures in all times. The universe seems to us now largely determinatively lawful excepting only some degree of apparently autonomous freedom of behavior within constraints of biological life. Human experience on earth seems to reflect a nature of regularity, cold neutrality and returning springs, and sometimes dangerous self-relevant forces. God seems beyond our rational certainty or refutation and to stay mostly outside of our affairs, leaving things to us, biology, and physical laws, except for our hopes and visions, and our obscure and variable inner intimations and deeper integrations. Science fiercely pursues this lawfulness in obedience to our deep inner quests for understanding and truth. Its grandest integrative views see all four levels--physical, biological, existential, and spiritual--as heirarchically nested and ultimately grounded in the physical. Yet, large important areas of uncertainty and ignorance still exist (for example, the existence and nature of other life in the cosmos), and many deep obscurities remain regarding the origins, natures, and self-relevant organizational patterns of most of these four realms. Our individual senses of autonomy and deeper intimations, and the widespread belief in these deeper sensibilities and in God are not altered by scientific discoveries, as powerful as these discoveries seem and as welcome as they are to all serious seekers of truth and understanding, both believers and sceptics.

The qualities attributed to God by humans, and human interpretations of His nature, will, and intimations, vary somewhat across cultures, as do the corresponding religions of the world. Yet, both show considerable common ground, particularly in showing strong emphases on considerate interpersonal ethical behavior, largely consonant with a compassionate humanist view, and on the desirability of higher meanings and aspirations in the life direction of individuals. In any case, a reality of God is not threatened by the weaknesses of human understanding. The grandest human senses of God in direct human experience are those of saints, prophets, and religious mystics which include central features of: peace beyond understanding, ultimate meaning, preciousness of all life, infinite sweetness, ultimate commonality of all life before death, all-inclusive benevolence, forgiveness, and love[17].

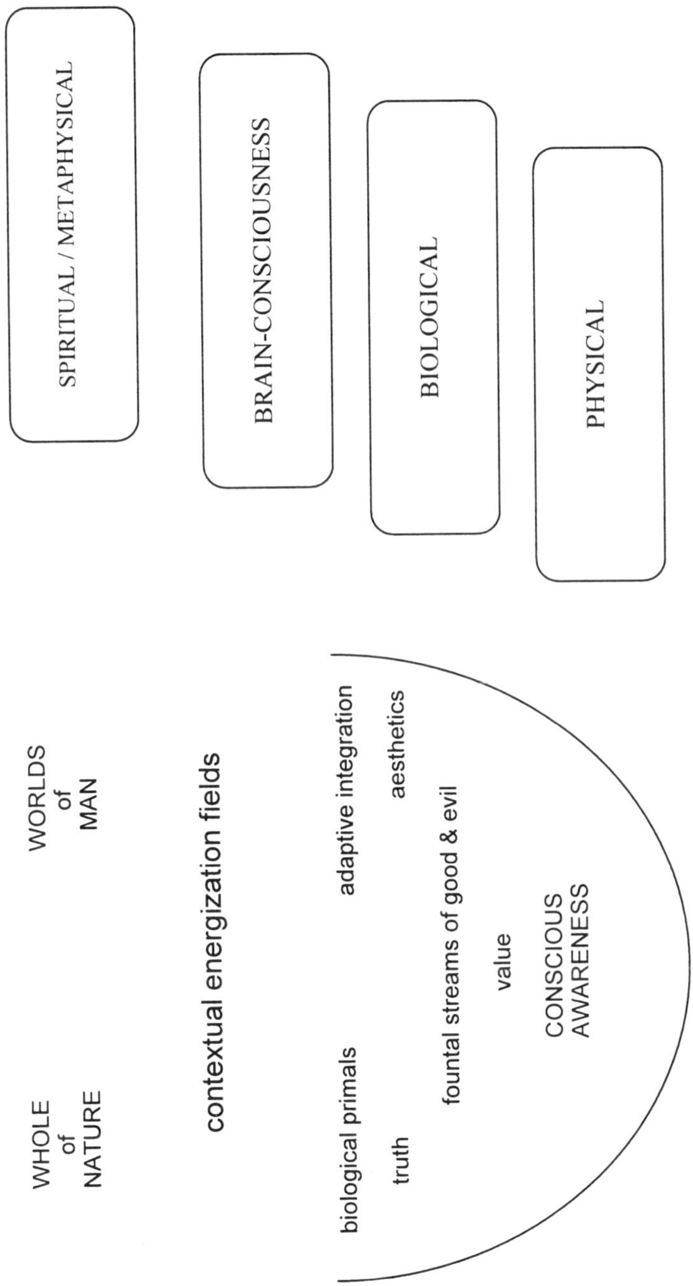

Fig. 16.1 Areas of metaphysical uncertainty.

All of this enhances individual human existence and human cooperation, and supports human survival. It also underscores the centrality of both consciousness and its freedom of will in the obscurities and mysteries of metaphysics, as well as in the guidance of human living and in human existence itself.

Endnotes

1. MacGregor1991,1993; Poznansaki2002
2. Festinger1959
3. Haack1993
4. Pollock2003, Greene1999, Walker2000
5. Beckley and Shapiro 1998
6. Ashbrook and Albright 1997
7. Tarnas1991, Kramer2002, MacGregor1995
8. Descartes1637, 1641; Kant1783-1790; Bergson 1908, 1919; Jackson1932; Freud1923, 1940
9. Tarnas1991, Kramer2002, Edwards KA, 1967, Solomon and Higgins KA, 1996
10. Nietzsche1885, Wittgenstein1945, Husserl1907, Ayer1956
11. Flanagan and Kim1995
12. Kierkegaard 1843, Sartre, 1943, Camus 1956, Kaufmann, 1956, Langiulli 1997
13. Tarnas1991, Kramer2002, Solomon and Higgins1996, Kuhn1962, 1969; Derrida1967, Lacan1966, Baudrillard1986
14. James1907, Haack1992, Levi-Strauss1977, Quine1961, Searle2000, Bergson 1919, 1920, Gadamer1981
15. post-modern thinking on knowing is criticized more explicitly in the last section of chapter ten
16. White1955
17. James 1902, St Teresa of Avila 1572

REFERENCES

Abeles, M 'Local Cortical Circuits', Springer-Verlag, Berlin, 1982.

Ackerman, D 'An alchemy of mind', 2004

Ashbrook, JB and CR Albright, 'The Humanizing Brain: Where Religion and Neuroscience Meet', Pilgrim Press, 1997.

Ayer, A 'The Problem of Knowledge', 1956

Baars, BJ, et al, 'Brain, conscious experience, and the observing self', Trends in Neurosciences, 26, 2003, 671-675.

Bach-y-Rita, P 'Nonsynaptic Diffusion Neurotransmission and Late Brain Reorganization (Demos Medical Publishers, New York, 1995.

Baudrillard, J 'America', 1986

Beckley (ed), B and D Shapiro, 'Uncontrollable Beauty --Towards a new Aesthetics', Allworth Pr, New York, 1998.

Bergson, H 'Matter and Memory', 5th ed, 1908. NM Paul and WS Palmer (tr), Zone Books, New York, 1991

Bergson, H 'L'Energie spirituelle', Paris, 1919; HW Carr (tr), 'Mind-Energy', New York, 1920.

Blanshard B 'The Nature of Thought', Macmillan Co, New York, 1939.

Bowlby, J 'Attachment and Loss', Vols 1 & 2, Basic Books, New York, 1969, 1973.

Bradford, HF 'Chemical Neurobiology', WH Freeman, 1986

von Braitenberg, H 'Cell assemblies in the cerebral cortex', pg 171 in R Heim & G Palm (eds) 'Theoretical approaches to complex systems', Springer, 1978

Brodmann, K 'Localisation in the Cerebral Cortex' (1909), LJ Garey (tr), Imperial College Press, 1999.

Camus, A 'The Fall', Vintage, NY, 1956

Campbell, J 'The Masks of God', Viking, New York, 1959-67.

Cannon, WB 'The Wisdom of the Body', Norton, 1944.

Cappa, SF 'Cognitive Neurology', Imperial College Press, 2001.

Chauvet, GA 'Theoretical systems in biology: hierarchical and functional integration', Pergamon, 1996.

Chauvet, GA 'On the Mathematical Integration of the Nervous Tissue Based on the S-propagator Formalism: I. Theory', J. Integrat. Neurosci., 1,1, 31-68, 2002.

Chomsky, N 'Language and Mind', Harcourt, Brace, Janovich, New York, 1972

Costalat, R & B Delord, 'Ephaptic Interactions Between Neurons', ch 14 in GN Reeke, et al, 'Modeling in the Neurosciences(2nd ed): From Biological Systems to Neuromemitic Robotics', Amsterdam, Harwood Academic Publishers, 2005.

Crick, F 'The astonishing hypothesis: the scientific search for the soul', Touchstone, Simon and Schuster

Crick, F and C Koch, 'a framework for consciouness', Nature Neuroscience, 6, 119-126, 2003.

Davies, P 'The Cosmic Blueprint',1988, 'God and the New Physics', 'The Mind of God', 1991, Simon and Schuster.

Davis, L and R Lorente de No, 'Contribution to the mathematical theory of electrotonus', Studies of Rockefeller Int Med Res, 131,442-496, 1947

Delacour , J (ed), 'The Memory System of the Brain', Imperial College Press, 1994

Dennett, D 'Consciousness explained', Back Bay, Little, Brown, 1991.

Derrida, J 'Of Grammatology', 1966; Johns Hopkins Univ Press, Baltimore, 1997

Descartes, R (1637,1641) Discourse on Method and Meditations (L.J. Lafleur, tr, 1960). Bobbs-Merrill Co, New York.

De Wall, F 'Tree of Origin: What primate behavior can tell us about human social evolution;, Harvard Univ Press, 2001

Eccles, JC 'The Physiology of Nerve Cells', John Hopkins Press, 1957.

Eccles, JC 'The Physiology of Synapses', Academic Press, 1964.

GM Edelman, 'Neural Darwinism: the theory of neuronal group selection', Basic Books, 1987.

Edelman, G 'The remembered present: a biological theory of consciousness', Basic Books, 1989.

Edelman, G 'Bright air, brilliant fire: on the matter of the mind', Basic Books, 1992.

Edelman, G 'Wider than the sky: Phenomenal Gift of Consciousness', Basic Books, 2004.

Edwards, P (ed), 'Encyclopedia of Philosophy', Macmillan Publ, New York, 1967

Ethology--see K Lorentz and Tinbergen.

Flanagan, O 'History of the Philosophy of Mind', pages 570-574 in T. Honderich (ed)'The Oxford Companion to Philosophy', Oxford Univ Presss, New York, 1995.

Fernandez-Armesto, F 'Truth', St Martin's Press, 1997.

Festinger, L 'The Theory of Cognitive Dissonance', Stanford University Press, Stanford, CA, 1957.

Freeman, WJ 'Waves, Pulses, and the Theory of Neural Masses', Prog Theor Biol, 1, pg 1-34, John Wiley, New York, 1972.

Freeman, WJ 'Neurodynamics: an exploration of in mesocopic brain dynamics', Springer, Berlin, 2000.

Freud, S 'New Introductory Lectures on Psychoanalysis', 1923; 'An Outline of Psychoanalysis', 1940, see J Strachey (trnsl), 1949, WW Norton; See also ' Basic Writings of Sigmund Freud', Random House, 1977.

Gadamer, H 'Reason in the Age of Science'; F Lawrence (tr), MIT Press, Cambridge, 1981

Gardner, HH 'The mind's new science', Basic Books,1985

Gerstein, G, DH Perkel, 'Simultaneously Recorded Trains of Action Potentials: Analysis and Functional Interpretation', Science, 164, 828-830, 1969.

Gerstein, GL, DH Perkel, KN Subramanian, 'Identification of functionally related neural assemblies', Brain Res, 140, 43-62, 1978

Globus, GG 'Towards a noncomputational cognitive neuroscience', J Cognit Neurosci, 4, 299-310,1992.

Gluck , MA, DE Rumelhardt, 'Neuroscience and connectionist theory', Lawrence Erlbaum, 1990;

Goldfinger, M 'Rallian Equivalent Cylinders Reconsidered', J Integrat Neurosci, 4,2, 227-265, 2005.

Gordon (ed), E 'Integrative neuroscience', Harwood, 2000

Gould, SJ 'The Structure of Evolutionary Theory', Harvard Univ Press, 2002

Greene, B 'The Elegant Universe', W. W. Norton, 1999.

Groves, PM, CJ Wilson, RJ MacGregor, 'Neuronal interactions revealed by statistical analysis of neuronal spike trains: the substantia nigra', in S Garttini, J Pujo, and R Samanin (eds), Interactions Between Putative Neurotransmitters in the Brain', Raven Press, New York, 1978.

Groves, PM 'A theory of the functional organization of the neostriatum and neostriatal control of voluntary movement', Brain Res, 5, 109-132, 1983.

Haack, S 'Evidence and Inquiry--towards reconstruction in Epistemology', Blackwell, 1993.

Hacking, I 'Representing and Intervening', Cambridge University Press, 1983.

Hameroff, SR 'Quantum Coherence in Microtubules: a Neural Basis for Emergent Consciousness?' J. Consc. Studies, 1, 98-118, 1994.

Harmon, LD, ER Lewis, 'Neural Modeling', Physiol Rev, 46, 513-591, 1966

Hasty, J, M Dolnik, V Rottschafer, J Collins, 'Synthetic gene network for entraining and amplifying cellular oscillations', Am Physical Soc, 88,14, 2002.

Hauser, M 'Wild minds'

Hebb, DO 'The organization of behavior', McGraw-Hill, 1949

Higgins, KM 'World Philosophy', the Teaching Company, Chantilly, VA, 2001.

Hodgkin, AL, AF Huxley, 'A quantitative description of membrane current and its application to conduction and excitation in nerve', J Physiol, 117, 500-544, 1952

Honderich, T (ed) 'The Oxford Companion to Philosophy', Oxford Univ Presss, New York, 1995.

Husserl, E 'The Idea of Phenomenology', 1907; L Hardy et al (tr), Kluwer Acad Pub, 1999

Jack, JJB, RW Tsien, and D Noble, 'Electric current flow in excitable cells', Clarendon Press, 1975

Jackson, JH 'Selected writings of John Hughlings Jackson', Hodder and Stoughton, 1932

James, W 'Principles of Psychology', McGraw-Hill, New York, 1890.

James, W 'The varieties of religious experience' (1902), New American Library, 1958.

James, W 'Pragmatism', 1907.

John, ER 'Switchboard vs statistical theories of learning and memory', Science, 177, 850-864,1977.

Journal of Integrative Neuroscience, http://www.worldscinet.com, World Scientific Publishing C, Lte, Singapore.

Joyce, J "Portrait of the Artist as a Young Man', 1916; Viking, New York, 1964.

Jung, CG, Collected Works, 18 Volumes, RFC Hull (tr), Pantheon (Bolligen Series), 1953-61.

Kant, E 'Prolegomena to any Future Metaphysics', 1783; Critique of Pure Reason', 1787; 'Critique of Practical Reason', 1788; 'Critique of Judgment', 1790;

Kaszniak, A (ed), 'Emotions, Qualia, and Consciousness', Imperial College Press, 2001

Kaufmann, W (ed), 'Existentialism from Dostoevsky to Sartre', Meridian books, New York 1956

Keller, EF 'The Century of the Gene', Harvard Univ Press, 2000

Kercel, SW, HJ Caulfied, P Bach-y-Rita, 'Bizarre Hierarchy of Brain Function', in KL Priddy, and PJ Angeline (eds), 'Intelligent Computing: Theory and Applications, Proc. SPIE, vol 5103, 2003.

Kierkegaard, S 'Fear and Trembling', 1843

Kim, J 'Problems of Mind' & 'The Mind-Body Problem', pages 574-580 in T. Honderich (ed)'The Oxford Companion to Philosophy', Oxford Univ Presss, New York, 1995.

King, BJ 'The Origins of Language: What nonhuman primates can tell us', School of American Research Press, Sante Fe, NM, 1999.

Koch, C 'The Quest for Consciousness', Roberts and Co, Hauppauge, NY, 2004.

Korn, H and P Faure, 'Is there chaos in the brain? II. Experimental evidence and related models', C. R. Acad. Sci,, 2003.

Kramer, L 'European Thought and Culture in the 19th and 20th Centuries' (two parts), The Teaching Co, Chantilly, VA, 2002.

Krishnamurti, J 'The Only Revolution', Harper & Row, New York, 1967.

Kruger, J 'Simultaneous individual recordings from many cerebral neurons', Rev Physiol Biochem Parmacol, 1983, 98,178-233

Kubie, LS 'A theoretical application to some neurological problems of the properties of excitation waves which move in closed circuits', Brain, 53, 166-177, 1930.

Kuhn, TS 'The Structure of Scientific Revolution', Chicago, 1962,1969.

Lacan, J 'Ecrits', 1966

Lakoff, G, M Johnson, 'Philosophy in the flesh', Basic Books, 1998.

Lakoff,G, RE Nunez, 'Where does mathematics come from', Basic Books, 2000.

Langiulli, N (ed), European Existentialism, Transaction Pub, New Brunswick, NJ, 1997

Lashley, KS 'the cerebral organization of experience', in FA Beach, et al (eds) 'The Neuropsychology of Lashley', McGraw-Hill, 1960.

Lehman, JJ, SK Jegtvig, 'Reactive attachment disorder: a preventable mental health disease/, J Chirop. Med, 3, 69-75, 2004.

Levi-Strauss, C 'Triste Tropiques'; J & D Weightman,(tr), Washington Square Press, New York, 1977

Lewis, ER, 'Neural Networks', Online Encyclopedia of Neuroscience, 2003

Lewis, ER, et al (eds) 'Diversity in Auditory Mechanics', World Scientific, Singapore, etc, 1997

Lewis, ER, 'Metamodeling'

Libet, B 'A testable field theory of mind-brain interactions'. J.Consc. Studies 1,119-126, 1994.

Lindsay, KA et al "Equivalent Cables—Analysis and Construction', ch s10-11 in GN Reeke, et al, 'Modeling in the Neurosciences(2nd ed): From Biological Systems to Neuromemitic Robotics', Amsterdam, Harwood Academic Publishers, 2005.

Luria, AR 'Higher Cortical functions in man', Basic Books, 1966

MacGregor, RJ ' A model for reticular-like networks: Ladder nets, recruitment fuses, and sustained responses', Brain Res, 41,345-363, 1972.

MacGregor RJ, ER Lewis, Neural Modeling, 1977, Plenum Press, New York

MacGregor, RJ Neural and Brain Modeling 1987, Academic Press, San Diego

MacGregor, RJ 'Sequential configuration model for firing patterns in local neural networks', Biol. Cyber, 65, 339-349, 1991.

MacGregor, RJ 'Theoretical Mechanics of Biological Neural Networks', 1993, Academic Press, Boston

MacGregor, RJ 'A theory of representation and embedding of memory traces in hippocampal networks', ch 18 in J MacGregor, Theoretical Mechanics of Biological Neural Networks, Academic Press, Boston, 1993.

MacGregor, RJ 'An Integrative Theory of Brain and Mind--The Inner Sensibilities', J Integrat Neurosci, 1,1, 23-29, 2002.

MacGregor, RJ Review of C Koch's 'The Quest for Consciousness' J Integrat Neurosci, 3,3, 359-362, 2004.

MacGregor,T Jolene, 'The Poetics of Loss', Ph D dissertation SUNY, Binghamton, 1995, UMI, Ann Arbor, MI.

MacLean, PD 'The triune brain, emotion, and scientific bias', in FO Schmitt (ed) 'The Neurosciences, a 2nd study program', Rockefeller Univ Press, 1970

Magistretti, PJ L Pellerin, 'Cellular mechanisms of brain energy metabolism and their relevance to functional brain imaging', Phil Trans R Soc (Lond), B354, 1155-1163, 1999.

Maslow, A Chapters 6 and 7 in 'Toward a Psychology of Being', 2nd ed, Van Norstrand, 1968.

McCulloch, WS, W Pitts, 'A logical calculus of the ideas immanent in nervous activity', Bull Math Biophys, 9,127, 147, 1943

Mithen, S 'the Prehistory of the Mind', Thames and Hudson, 1994.
Moore, GP, DH Perkel, JP Segundo, 'Statistical Analysis and Functional Interpretation of Neuronal Spike Data', Ann Rev Physiol, 28,493-522, 1966.
Nietzsche, F 'Beyond Good and Evil', 1886; W Kaufman (tr), Vintage, New York, 1966.
Nobak, CR, NL Stromiger, & RJ Demarest, 'The Human Nervous System', Leas & Febiger, Philadelphia, 1991
Nunez, PL 'Electrical Fields of the Brain', Oxford Univ Press, New York, 1981.
Penfield, W 'The Mystery of the Mind', Princeton Univ Press, 1975.
Penrose, R 'The Emperor's New Mind', Oxford Univ Press, 1989.
Penrose, R 'Shadows of the mind: a search for the missing science of consciousness', Oxford Univ Press, 1994.
Pinker, S 'How the Mind Works', Norton, 1995
Pollack, S 'Particle Physics for Non-Physicists' The Teaching Co, Chantilly, VA, 2003
Popper, KR 'The Logic of Scientific Discovery', Hutchinson, London, 1959
Popper, KR, BIB Lindahl, A Arhem 'A discussion of the mind-brain problem'. Theor. Med. 14, 167-180, 1993.
Porter, R 'A mind and its meanings',(a review of F Crick: 'The Astonishing Hypothesis'), Times Literary Supplement, Sept 16, 1994.
Poznanski, RR (ed), 'Modeling in the Neurosciences (1rst ed): From Ionic Channels to Neural Networks', Amsterdam, Harwood Academic Publishers, 1999..
Poznanski, RR (ed), 'Biophysical neural networks: Foundations of Integrative Neuroscience', Mary Ann Liebert, 2001
Poznanski, RR 'Introduction to Integrative Neuroscience', ch 1 in R Poznanski (ed) 'Biophysical Neural Networks: Foundations of Integrative Neuroscience', Mary Ann Liebert, 2001.
Poznanski, RR 'Towards an Integrative Theory of Cognition', J Integrat Neurosci, 1,2,145-156, 2002.
Quine, W 'From a Logical Point of View', 2nd ed, Harper & Row, New York, 1961
Rall, W 'Core conductor theory and cable properties of neurons', in JM Brookhart, et al (eds) 'Handbook of Physiology, Rev ed, sect 1, vol 1, 39-97, Am Physiol. Soc, Baltimore Md, 1977
Ramon Y Cajal, S 'Histologie du system nervoux', 1911.
Ramon-Moliner,E 'The Conscious States of Matter', 1994.
Redhead, M 'From Physics to Metaphysics', Cambridge University Press, 1995.
Reeke, GN et al, 'Modeling in the Neurosciences(2nd ed): From Biological Systems to Neuromemitic Robotics', Amsterdam, Harwood Academic Publishers, 2005.
Reggia, JA, E Ruppin, and RS Berndt, 'Neural Modeling of Brain and Cognitive Disorders', Imperial College Press, 1996.
Ridley, M 'Genome', Harper Collins, 1999
Rosenblatt, F 'Principles of Neurodynamics', Spartan Books, Washington DC, 1962
Rosenfield, I 'The strange, familiar, and forgotten: an anatomy of consciousness', Vintage Press

St. Teresa of Avila, 'Interior Castle',1572, (E. Allison Peers, tr), Image, NY, 1961.
Sarnat, HB MG Netsky, 'Evolution of the Nervous System', Oxford, New York, 1974
Sartre, J 'Being and Nothingness', 1943
Scheibel, ME AB Scheibel, 'Structural substrates for integrative patterns in the brain stem reticular core', HH Jasper et al (eds) 'Reticular Formation of the Brain', Little, Brown and Co, Boston, 31-36, 1958.
Schore, AN 'Dysregulation of the right brain: a fundamental mechanism of traumatic attachment and the psychopathogenesis of posttraumatic stress disorder', Austal and N Zeal J of Psychiatry,36,9-30,2002.
Searle, JR 'Minds, Brains, and Science', Harvard, 1984.
Searle, JR 'The Rediscovery of the Mind', MIT Press, Cambridge, 1992.
Searle, JR 'The mystery of Consciousness', I and II', New York Review of Books, Nov 2 and 16, 1995.
Searle, JR 'Consciousness and the Philosophers', New York Review of Books, Mar 6, 1997.
Searle, JR 'Consciousness', Annu. Rev. Neurosci., 23,557-578, 2000.
Sherrington, C 'The Integrative Activity of the Nervous System', Yale Univ Press, 1906
Solomon, RC, KM Higgins, 'A Short History of Philosophy', Oxford University Press, New York, 1996.
Stenger, VJ 'the Unconscious Quantum', Prometheus Books, Amherst, NY, 1995.
Sykes, B 'The Seven Daughters of Eve: The science that reveals our genetic ancestry', WW Norton, 2001
Taddei-Ferretti, C, C Musio, 'Neuronal Bases and Psychological Aspects of Consciousness', Imperial College Press, 1999.
Tarnas, R 'The Passion of the Western Mind', Ballantine Books, 1991.
Taylor, C 'Sources of the Self', Cambridge, 1989.
Tudge, C 'The Variety of Life', Oxford, 2000
Turing, A 'Computing Machinery and Intelligence', Mind, 1950
Walker, EH 'the Physics of Consciousness', Perseus, 2000.
Watson, JA 'The human DNA molecule', 2004
White, A 'The Warfare of Science with Theology in Christendom', George Brazillier, New York, 1955
Wegner, MD 'The illusion of Conscious Will', MIT Press, Cambridge, 2003
Weinberg, S New York Review of Books, 6/12/97,of books by T Ferris,A Guth,M Rees.
Wiener, N 'Cybernetics: control and communication in the animal and the machine', 2nd ed, MIT Press, 1961
Wilson, CJ, HT Chang, and ST Kitai, 'Origins of postsynaptic potentials evoked in identified rat neostraital neurons by stimulation in substantia nigra', Exp Brain Res, 8, 157-162, 1982.
Wilson, E 'Consilience--the unity of knowledge'.
Wittgenstein, L 'Philosophical Investigations', 1945; G Anscomb (tr) , Macmillan, 1958
Yew, DT, WH Kwong, & MC Yu, 'Basic Neuroanataomy', Imperial College Press, 1997
Yusuf, HKM 'Understanding the Brain and its Development: A Chemical Approach', Imperial College Press, 1992.

DETAILED TABLE OF CONTENTS

Ch 1 Introduction and Overview 1

General

Approach 2

Consciousness and brain : A multicapacitied monism 2

The obscurities of consciousness 4

The grand unifying vision of science 7

Overview 8

Personal Proclivities 9

Terminology 10

Endnotes 12

Ch 2 An Integrative View of the Brain and Conscious Experience 15

General

Individual Components of the Inner Sensibilities 17

Capacities of the biological primals (Table 2.1) 18

Capacities of adaptive integration 19

Capacities of valuative representation 20

Generalized aesthetic perception 21

Apprehension-cognition 21

Value, good, and bad 23

The valuative representations as a whole 24

Motivation and Inner Directive Causes, Energization 25

The Significance of the Valuative Representations 27

The Justification and Obscurities of Physical-Experiential Organicity 29

Endnotes 30

Ch 3 The Structure of Human Value 31

General

Five Root-Types of Value (Fig. 3.1) 34

Seven Domains of Value (Table 3.1) 37

The Peak Experiences 40

Comment 40

Endnotes 41

Ch 4 The Inner Sensibilities and Neurobiology 43

Higher Integrative Plasticity in Humans 43

General

Open-endedness and adaptivity 45

Self-direction, autonomy, consciousness, and evolution 46

Evolution and Brain Organization 47

The tyranny of survivability 50

Functional Topography of the Brain 50

Operational Organization of the Brain 52
Motivation and human causation 52
The hierarchical stratification of brain variables [Table 4.1(a)] 53
Consciousness and Neurobiology 55
The Inner Sensibilities and Neurobiology 57
General
Organic cooperation of the inner sensibilities 58
The neurobiological topography of the inner sensibilities 60
Neurobiological foundations of the capacities of the inner sensibilities [Table 4.1(b)] 60
Value and neurobiology 62

Ch 5 Consciousness and Theoretical Neuroscience 69
Introduction and Current Literature 69
Neural Networks: Gerald Edelman's view of Neuronal Group Dynamics and Consciousness 71
Integrative Neuroscience: Multilevel Hierarchy of Brain Physiology (Fig. 5.1) 73
Neurophysiology and Physical processes 78
Neuroelectric signals 79
Ionic fluxes 79
Mechanisms of neuroelectric signaling 80
The neurochemical dimension 81
Neural theory and neurophysiological processes 82
The Computer Metaphor for Brain Function 83
Endnotes 87

Ch 6 Consciousness, Physics, and Neurobiology 91
Texture, Interfaces, Energization, and Nature of Consciousness 91
Texture 91
Interfaces 92
Energization 94
The fundamental nature of consciousness 95
Conscious Experience and Physical Science 96
The grand unifying view of physical science 96
Molecular biology and life, and consciousness and autonomy 97
Theoretical physics: Quantum mechanics and string theory 99
Two Speculative Views of Quantum Mechanical Consciousness 101
Placing Conscious Autonomy in Physical Science 104
Interfacing quantum mechanical and classical physics 105
The nature of quantum mechanical uncertainty 107
Overall operation of conscious autonomy 108
Types and principles of possible quantum mechanical effects in brain 108
Concluding Assessment of Quantum Mechanics and Consciousness 112
Does the brain do it all? 116
What Are We? Fundamental Views of the Nature of Consciousness 116

Endnotes 119

Ch 7 A Functional Theory of Consciousness and its Relations in Brain 121
General
A Functional Theory of Consciousness 122
The flowering of consciousness 122
The phenomena of consciousness 123
Fundamental tenets of the theory 124
Characteristics of The Theory 127
General
Summary of the processes of this theory of consciousness 129
Localization of Consciousness in the Brain 129
Possible neurobiological correlates 130
Glial cells and brain metabolism 132
Broad interpretations regarding localizations of consciousness 132
Comments 134
Endnotes 135

Ch 8 Consciousness, Inner Constructions, and Language 137
The Range and Nature of Inner Constructions 137
Language 138
The structural Nature and Place of Language in Neurobiology 139
The outer conscious-related face of language 139
The inner neurobiological roots of language (Fig. 8.1) 140
The Conscious and Cognitive Use of Language 143
Endnotes 143

Ch 9 First Principles of Human Awareness and Apprehension 145
General
Pre-Analytic Apprehension and Inherent
and Foundationo-Rational Knowing 146
General
Consensus, coherence, correspondence, and inherent
and foundationo-rational truth 147
First Principles of Human Existence and Apprehension 148
Intrinsic placement 148
Mind-body question 149
Organic functionalism 149
Universal human perspective 150
Foundations of human knowing 151
Comment on first principles 152
Pre-analytic knowing and language 153
Energization and the Inner Sensibilities 153
The Contexts of Things Human 154
Our existence in time 154

Our place in space — the radiality of human existence	155
Primordial differentiations	156
Further Remarks on Foundationo-Rational Knowing	156
Endnotes	157
Ch 10 Knowing and Neurobiology	159
General	
Inner Patterns and Knowing	159
A Typology of Knowing (Table 10.1)	163
Knowing and Contemporary Thought	165
Endnotes	168
Ch 11 The Larger Nature and Setting of Value	169
General	
The Essential Nature and Structure of Value	169
General	
Intrinsic values	172
Sociopersonal values	172
Valenced thoughts	173
Ethics and Moral Philosophy	175
General	
The philosophies of religion and value	176
Wisdom [Table 11.1(a)]	176
Placing the Values of the World's Religions [Tables 11.1(b) and (c)]	180
Endnotes	182
Ch 12 A Way of Placing Things	189
General (Fig. 12.1)	
Qualities of This Map of the Whole of Things	189
Realms and Headings of the Map	192
First- and second-level realms of the map [Tables 12.1(a) and (b)]	192
Comments	196
Endnotes	196
Ch 13 The Universal Human Condition	199
General	
Fundamentals of the Human Condition	200
Individual existence in itself — foundations	200
Individual existence in terms of surrounding reality — foundations	201
Human Nature	202
General	
Snapshot of human nature	204
Bald skeletal statement of human nature	205
Science and Nature	205
Obscurities and mysteries	206

The physical universe 206
The biological universe 207
The sociobehavioral universe 207
Conscious experience and apprehension 207
Worlds of Man 208
Broad Summary of the Human Condition [Tables 13.1(a) and (b)] 210

Ch 14 Fountal Streams of Good and Evil 217
Fountal Streams of Good and Evil (Fig. 14.1) 217
Empathy and Neural Development 220
Karma 223
The Good Life and Morality — Scope and Contexts 224
The Good Life and Morality — Pith 228
General
Morality 232
Morality, Darwinian Selection, Cultural Evolution, and Species Survival 233
Closing Comments 237
Endnotes 238

Ch 15 Contextual Energization Fields 239
General
Elemental Contextual Energization Fields (Table 15.1) 240
The individual contextual energization Fields 240
Fundamental Contextual Energizations of Personal Living (Fig. 15.1) 244

Ch 16 Implications and Interpretations 247
Particular Implications of This Work 247
Knowing, neurobiology, and epistemology 247
Neuroscience 247
Epistemology 248
Foundations and function of consciousness 248
Theoretical physics 248
Brain organization 249
The fuller inner sensibilities and its qualities 249
Fuller ambient contexts 250
Value, good and evil, morality 250
The Inner Sensibilities and Academic Philosophy 250
General
Twentieth-century philosophy 251
Critique of the four main movements of twentieth-century philosophy 253
The inner sensibilities and academic philosophy 254
The Inner Sensibilities and The Human Potential 256
Division and overreduction 256
The inner sensibilities and common ground 257
The Ambient Contexts of Things Human 259

The direction of human living 259
A positive view of man 259
The central human problems of our times 261
Spirit, Metaphysics, Science, Religion, and the Inner Sensibilities 263
Monism, determinism and free-will, spirit, religion, and science 263
Metaphysics, spirit, science, and religion (Fig. 16.1) 266